Cattedra Galileiana
Pisa 2000

Tomas Björk

A Geometric View
of the Term Structure
# of Interest Rates

ISBN: 978-88-7642-241-6

Lecture Notes written by

Irene Crimaldi

CONTENTS

Preface . . . . Pag.   1

Chapter 1 - Introduction and Basic Knowledge . . . . ”   3

  1.1. – Preliminaries . . . . ”   3

  1.2. – Short rate models . . . . ”   7

  1.3. – Affine term structure . . . . ”   11

  1.4. – Inverting the yield curve . . . . ”   11

  1.5. – Heath - Jarrow - Morton . . . . ”   13

  1.6. – Musiela parameterization . . . . ”   14

Chapter 2 - The Model . . . . ”   15

Chapter 3 - Linear Realizations of Interest Rate Models . . . . . ”   17

  3.1. – Assumptions . . . . ”   17

  3.2. – Systems theory and realization theory . . . . ”   18

  3.3. – Realizations of interest rate models . . . . ”   21

  3.4. – Existence of finite dimensional realizations . . . . ”   24

3.5. – Minimal realizations . . . . "   27

3.6. – Economic interpretation of the state space . . . . "   29

**Chapter 4 - Interest Rate Dynamics and Consistent Forward Rate
Curves** . . . . "   31

4.1. – The formal problem statement . . . . "   33

4.2. – The deterministic finite dimensional case . . . . "   36

4.3. – The stochastic finite dimensional case . . . . "   37

4.4. – Invariant forward curve manifold . . . . "   39

4.5. – Concrete examples . . . . "   40

4.6. – New forward curve families . . . . "   42

4.7. – The Filipoviĉ approach . . . . "   44

Chapter 5 - Finite Dimensional Realizations of Forward Rate Models "   47

5.1. – Some differential geometry . . . . "   47

5.2. – The space . . . . "   54

5.3. – Finite dimensional realizations . . . . "   56

5.4. – Deterministic volatility . . . . "   57

5.5. – A concrete example . . . . "   58

5.6. – Constant direction volatility . . . . "   61

5.7. – Short rate realizations . . . . "   63

References . . . . "   67

Preface

This set of lecture notes is the outcome of a lecture series, given in April 2000 by the author while holding the "Cattedra Galileiana" at Scuola Normale Superiore in Pisa. The purpose of the lectures was to give an overview of some recent work concerning structural properties of the evolution of the forward rate curve in an arbitrage free bond market. The audience consisted of doctorate students and faculty from the Scuola Normale. I would like to convey my thanks to the Scuola Normale for inviting me, and to Professors Aubin, Da Prato, Zabczyk and Pratelli for a large number of interesting discussions. Professor Pratelli was a truly splendid host, and I am also very grateful to everyone in the audience for providing a very stimulating atmosphere.

Irene Crimaldi had the hard work of transforming my lectures into their present form. She has done a magnificent work and I am truly grateful.

Stockholm, December 6, 2000.

Tomas Björk

CHAPTER I

Introduction and Basic Knowledge

The purpose of these lectures is to give an overview of some recent work concerning structural properties of the evolution of the forward rate curve in an arbitrage free bond market. The main problems to be discussed are as follows:

- When is a given forward rate model consistent with a given family of forward rate curves?

- When can the inherently infinite dimensional forward rate process be realized by means of a finite dimensional state space model?

In this chapter we sum up some basic results of interest rate theory. We suppose the reader to be familiar with this theory and we refer to [1] for more details.

1.1. – Preliminaries

We take as given a filtered complete probability space $(\Omega, \mathcal{F}, P, \underline{\mathcal{F}})$, where the filtration $\underline{\mathcal{F}}$ satisfies the usual hypotheses, and a vector process $S(t) = [S_0(t), \ldots, S_N(t)]$, where $S_i(t)$ is interpreted as the price of one unit of asset number i at time t. We suppose that the processes $S_i(t)$

are adapted and defined by a system of stochastic differential equations, driven by a d-dimensional Wiener process.

In the sequel we need the following definitions:

DEFINITION 1.1.1. *A **portfolio** is a predictable vector process*

$$h(t) = [h_0(t), \ldots, h_N(t)],$$

where $h_i(t)$ is interpreted as the number of units of asset i at time t.

DEFINITION 1.1.2. *The **value process** corresponding to a portfolio h is defined by*

$$V^h(t) = \sum_{i=0}^{N} h_i(t) S_i(t).$$

DEFINITION 1.1.3. *A portfolio h is said to be **self financing** if we have*

$$dV^h(t) = \sum_{i=0}^{N} h_i(t)\, dS_i(t),$$

where the processes h_i are supposed such that the equality above has sense.

DEFINITION 1.1.4. *A portfolio h is an **arbitrage strategy** if h is a self financing portfolio such that*

(i) $V^h(0) = 0$,

(ii) $P\left(V^h(t) \geq 0\right) = 1$ *for $0 \leq t \leq T$,*

(iii) $P\left(V^h(T) > 0\right) > 0$.

*If no arbitrage portfolios exist for any $T \in \mathbb{R}_+$, we say that the model is **free of arbitrage** or **arbitrage free**.*

As can be easily seen from the definition, when we have an arbitrage portfolio, we have an opportunity to make money without risk. Then the first problem is to investigate when a given model is free of arbitrage. The main result in this sense is the following:

THEOREM 1.1.5. *Let us assume that the process S_0 is a.s. strictly positive. Then the market is arbitrage free if and only if there exists a probability measure $Q \sim P$ such that all discounted price processes are Q-martingales, that is*

$$Z(t) = \frac{S(t)}{S_0(t)} = [1, Z_1(t), \ldots, Z_N(t)]$$

is a Q-martingale.

We say that the probability measure Q above is a **martingale measure**.

REMARK 1.1.6. The Theorem 1.1.5 is informally true, but in reality the exact theorem is more complicated. It is easy to see that the existence of a martingale measure Q implies the absence of Q-**admissible** arbitrage portfolios, that is arbitrage portfolios such that the discounted value process is a Q-martingale. On the other hand, the converse is more complicated and we refer to [7] for a precise formulation.

In the sequel, when we speak of absence of arbitrage, we will really mean that there exists a martingale measure.

The **numeraire** price S_0 can be chosen arbitrarily. We choose the riskless asset, that is

$$S_0(t) = B(t),$$

where

$$\begin{cases} dB(t) = R(t)B(t)\, dt \\ B(0) = 1 \end{cases}$$

with R a given adapted stochastic process.

The process B is termed money account and describes a bank with short rate R.

Now let us consider the **"pricing problem"**.

DEFINITION 1.1.7. *A contingent **T-claim** is any random variable X in $L_+^0(\Omega, \mathcal{F}_T, P)$, that is any non-negative $\mathcal{F}_T$-measurable random variable.*

A contingent claim can be interpreted as a contract which specifies that the stochastic amount, X, of money is to be paid to the holder of the contract at time T.

The price problem consists of determining a "reasonable" price process $\pi(t; X)$ for a fixed contingent T-claim X.

One approach to this problem is to demand that the vector process $[S_0(t), \ldots, S_N(t), \pi(t; X)]$ is free of arbitrage. To this end, we assume that there exists a martingale measure Q such that the discounted claim $X/S_0(T)$ is Q-integrable and we impose that the process $\pi(t; X)/S_0(t)$ is a Q-martingale. We thus have the following *pricing formula*:

$$\pi(t; X) = B(t)\, E^Q\left[\frac{X}{B(T)}\,\Big|\,\mathcal{F}_t\right] = E^Q\left[e^{-\int_t^T R(s)\, ds} \times X \mid \mathcal{F}_t\right].$$

We say that the process $\pi(t; X)$, defined above, is an **arbitrage free price process** for X and this way to price contingent claims is said **"martingale pricing"**.

Note that, for a fixed T-claim X, different choices of Q will generally produce different price processes for X.

We have also an other approach to the pricing problem.

DEFINITION 1.1.8. *A portfolio h is a* **hedge** *against a T-claim X if h is a self financing portfolio such that*

$$V^h(T) = X \qquad P - \text{a.s.}$$

We also say that the portfolio h **replicates** *X or X is* **replicated** *by h.*

Then, if h replicates X, a natural way of pricing X is to set

$$\pi(t; X) = V^h(t).$$

It can be easily proved that, if Q is a martingale measure and if the replicating portfolio h is Q-admissible, then the vector process $\pi(t; X)/S_0(t)$ is a Q-martingale and we have

$$\pi(t; X) = V^h(t) = E^Q \left[e^{-\int_t^T R(s)\,ds} \times X \mid \mathcal{F}_t \right].$$

Thus, in an arbitrage free market, different choices of admissible replicating portfolios will produce the same price.

DEFINITION 1.1.9. *The market is said to be* **complete** *if every contingent claim X can be replicated.*

In the sequel, the word *"complete"* will actually be used informally, that is in the sense that some reasonably large subset of the set of contingent claims can be replicated.

We have the following result:

THEOREM 1.1.10. *The market is complete if and only if the martingale measure Q (if it exists) is unique.*

This result, modulo some suitable technical conditions, is a consequence of Yor-Jacod Lemma. More details can be found in [10].

Now we conclude this section with a rule of thumb for quickly determining whether a certain model is complete and/or free of arbitrage.

META-THEOREM 1.1.11. *Assume that N is the number of underlying assets in the model* **excluding** *the numeraire asset, and S the number of independent sources of randomness. In the generic case we then have the following relations:*

(i) *The market is arbitrage free if and only if $S \geq N$.*

(ii) *The market is complete if and only if $S \leq N$.*

(iii) *The market is arbitrage free and complete if and only if $S = N$.*

This is a purely heuristic result but, in concrete cases, it can be given a precise formulation and a precise proof.

EXAMPLE 1.1.12. (**Black - Scholes Model**)
Let us consider the following model:

$$\begin{cases} \mathrm{d}S = \alpha S \,\mathrm{d}t + \sigma S \,\mathrm{d}W \\ \mathrm{d}B = RB \,\mathrm{d}t. \end{cases}$$

(The filtration $\mathcal{F}$ is defined as the usual enlargement of the filtration generated by the Wiener process W.)

We have $N = S = 1$ and it can be proved that, in this case, the market is arbitrage free and complete.

1.2. – Short rate models

Let us start with the following definition:

DEFINITION 1.2.1. *A* **zero coupon bond** *with maturity date T (also called a T-bond) is a contract which pays $1\$$ to the holder at time of maturity T.*

We now want to study the market of zero coupon bonds. At this end, let us denote with $p(t, T)$ the price process at time t of a bond with maturity date T.

We suppose that there exists a market for T-bonds for every $T \geq 0$. Moreover we take as given a filtered complete probability space $(\Omega, \mathcal{F}, P, \mathcal{F})$, where $\mathcal{F}$ is the usual enlargement of the filtration generated by a P-Wiener process $\widetilde{W}$, and, for every fixed t, we make the assumption that $p(t, T)$ is, P-a.s., continuously differentiable in the T-variable.

Now we can give the following definitions:

DEFINITION 1.2.2. *The* (*instantaneous*) **forward rate** *with maturity T, contracted at t is defined by*

$$f(t, T) = -\frac{\partial \log p(t, T)}{\partial T}.$$

DEFINITION 1.2.3. *The* (*instantaneous*) **short rate** *at time t is defined by*

$$R(t) = f(t, t).$$

We assume that the P-dynamics of the short rate $R(t)$ is given by

$$dR(t) = \mu(t, R(t)) \, dt + \sigma(t, R(t)) \, d\widetilde{W}(t),$$

where $\mu(t, r)$ and $\sigma(t, r)$ are real valued functions which are supposed regular enough in order to have a unique strong solution.

Further we consider the **money account process** defined by

$$B(t) = e^{\int_0^t R(s) \, ds}.$$

The main problems to be solved are the following:

1. Determine the **term structure**, that is the structure of $\{p(t, T) \, ; \, 0 \leq t \leq T \, , \, T \geq 0\}$ in an arbitrage free bond market.

2. Determine arbitrage free prices of other interest rate derivatives.

Using the results of the previous section, we can observe that the P-dynamics of R and the requirement of an arbitrage free bond market are not sufficient to determine the bond prices. In fact, if the money-account B is the only exogenously given asset, every $Q \sim P$ is a martingale measure and so the market is not complete. Besides, using the notations of Meta-Theorem 1.1.11, we have $N = 0$ (no risky asset) and $S = 1$ (one source of randomness, $\widetilde{W}$) and so we find again that the market is incomplete.

Summing up, there is not a unique price for a particular T-bond. However, in order to avoid arbitrage, bonds of different maturities have to satisfy internal consistency relations: if we take a T_0-bond as a "benchmark bond" (i.e. as given in the market) then other bonds can be priced in terms of the market price of the benchmark bond.

Let us suppose that the price process of a T-bond is of the form

$$p(t, T) = F(t, R(t), T) = F^T(t, R(t)),$$

where the pricing function $F(t, r, T)$ is a strictly positive smooth function of the three variables.

Using Itô's formula we obtain the following T-bond dynamics:

(1.1) $$\mathrm{d}\, F^T = F^T m_T \, \mathrm{d}t + F^T v_T \, \mathrm{d}\widetilde{W},$$

where

(1.2) $$\begin{cases} m_T = \dfrac{F_t^T + \mu F_r^T + \frac{1}{2}\sigma^2 F_{rr}^T}{F^T} \\[2mm] v_T = \dfrac{\sigma F_r^T}{F^T}. \end{cases}$$

Then, by a very easy argument, we can find out the relations that must hold between pricing functions for different maturities in an arbitrage free market. More precisely, we have the following result:

THEOREM 1.2.4. *There exists a universal process $\lambda(t)$ such that*

(1.3) $$\frac{m_T(t) - R(t)}{v_T(t)} = \lambda(t), \qquad \text{for all } t \text{ and } T.$$

The process λ can be interpreted as "the risk premium per unit of volatility" and it is termed **"the market price of risk"**.

From formulae (1.2) and (1.3), we obtain the following **"Term Structure Equation"**:

(1.4) $$\begin{cases} F_t^T + \{\mu - \lambda\sigma\} F_r^T + \frac{1}{2}\sigma^2 F_{rr}^T - r F^T = 0 \\[2mm] F^T(T, r) = 1. \end{cases}$$

By the Feynman-Kač formula, we find out the following representation of the solution of equation (1.4):

$$F(t, r, T) = E_{t,r}^Q \left[e^{-\int_t^T R(s)\, \mathrm{d}s} \right],$$

where the Q-dynamics for the short rate is given by

(1.5) $$\begin{cases} \mathrm{d}R(s) = \{\mu - \lambda\sigma\}\mathrm{d}s + \sigma\, \mathrm{d}W(s) \\[2mm] R(t) = r. \end{cases}$$

(We have indexed the expectation operator by (t, r) in order to emphasize that the expected value is to be taken given the initial value $R(t) = r$.)

We can easily prove that the above measure Q is a martingale measure for the bond market and we can note that specifying the market price of risk is equivalent to specifying the martingale measure Q. So we can model the short rate directly under a *fixed martingale measure Q* and we have the following result:

THEOREM 1.2.5. *Let us assume that the Q-dynamics of the short rate is*

$$dR = \mu(t, R)dt + \sigma(t, R) dW.$$

Let X be a T-claim. Then the arbitrage free price process $\pi(t; X)$ is given by

$$\pi(t; X) = E^Q\left[e^{-\int_t^T R(s)\, ds} \times X \mid \mathcal{F}_t\right].$$

In particular, in the case $X = \Phi(R(T))$ (with Φ sufficiently regular), the process $\pi(t; X)$ is given by

$$\pi(t; X) = F(t, R(t)),$$

where F solves the Term Structure Equation

(1.6)
$$\begin{cases} F_t + \mu\, F_r + \frac{1}{2}\sigma^2\, F_{rr} - r\, F = 0 \\ F(T, r) = \Phi(r) \end{cases}$$

and so it has the representation

$$F(t, r) = E_{t,r}^Q\left[e^{-\int_t^T R(s)\, ds} \Phi(R(T))\right].$$

We conclude this section giving a list of some martingale models (i.e. models under a fixed martingale measure Q) for the short rate process:

1. *Vasiček*
$$dR = (b - aR)\, dt + \sigma\, dW,$$

2. *Cox-Ingersoll-Ross*
$$dR = (b - aR)\, dt + \sigma\sqrt{R}\, dW,$$

3. *Dothan*
$$dR = aR\, dt + \sigma R\, dW,$$

4. *Black-Derman-Toy*
$$dR = a(t)R\, dt + \sigma(t)R\, dW,$$

5. *Ho-Lee*
$$dR = a(t)\, dt + \sigma\, dW,$$

6. *Hull-White (extended Vasiček)*
$$dR = \{\Phi(t) - aR\}\, dt + \sigma\, dW.$$

1.3. – Affine term structure

As we can see from Theorem 1.2.5, we need an interest rate model for which it is easy to compute the solution of equation (1.6).

The main result in this direction is about the existence of a so called "affine term structure".

DEFINITION 1.3.1. *We have an* **affine term structure** *if the bond prices have the form*

$$(1.7) \qquad p(t, T) = F(t, R(t), T) = e^{A(t,T) - B(t,T) R(t)},$$

where A and B are deterministic functions.

Then we have the following result:

THEOREM 1.3.2. *With the same notation of Theorem 1.2.5, let us assume that μ and σ are of the form*

$$\begin{cases} \mu(t, r) = \alpha(t)r + \beta(t) \\ \sigma^2(t, r) = \gamma(t)r + \delta(t). \end{cases}$$

Then the model admits an affine term structure of the form (1.7), where A and B satisfy the system

$$\begin{cases} B_t(t, T) = -\alpha(t)B(t, T) + \frac{1}{2}\gamma(t)B^2(t, T) - 1 \\ B(T, T) = 0 \end{cases}$$

and

$$\begin{cases} A_t(t, T) = \beta(t)B(t, T) - \frac{1}{2}\delta(t)B^2(t, T) \\ A(T, T) = 0. \end{cases}$$

1.4. – Inverting the yield curve

In this section, we want to introduce the problem of "calibrating the model to data" or "fitting the yield curve".

In order to fit an interest rate model to real data, we can proceed in the following way:

1. Fix a Q-dynamics with a parameter vector α:

$$dR = \mu(t, R; \alpha)dt + \sigma(t, R; \alpha)\,dW.$$

2. Solve, for all fixed times of maturity T, the term structure equation in order to obtain the theoretical bond price $p(t, T; \alpha)$.

3. Choose the parameters vector α in such a way that the theoretical term structure $\{p(0, T; \alpha); T \geq 0\}$ fits, as close as possible, the empirical term structure $\{p^*(0, T); T \geq 0\}$ observed on the market at time $t = 0$.

Thus, following the previous procedure, we want to find a parameter vector α^* such that α^* solves the equation

(1.8) $p(0, T; \alpha) = p^*(0, T),$ for all $T \geq 0$.

However we can observe that (1.8) is an infinite dimensional system of equations (one for each T) and so, if we want a good fit between observed and theoretical bond prices, we have to work with models containing an infinite parameter vector (for example, Ho-Lee or Hull-White models).

Summing up, the main advantages of the models based on the short rate are the following:

1. Specifying R as the solution of a stochastic differential equation allows us to use Markov process theory and we may work within a PDE framework.

2. It is often possible to obtain analytical expressions for bond prices and derivatives.

Instead, the main drawbacks are:

1. It is hard to obtain a realistic volatility structure for the forward rates without using a very complicated model.

2. Inverting the yield curve can be hard: the more realistic the model is, the more difficult the calibration described above becomes.

3. One factor models implies perfect correlation along the yield curve.

1.5. – Heath - Jarrow - Morton

In order to avoid the problems described in the previous section, Heath, Jarrow and Morton propose to model the dynamics for the entire forward rate curve and use the observed forward rate curve as boundary value at $t = 0$. More precisely, it is assumed that, for each fixed T, the forward rate $f(t, T)$ has a dynamics, under the objective measure P, given by

$$(1.9) \qquad df(t, T) = \tilde{\alpha}(t, T)\, dt + \sigma(t, T)\, d\widetilde{W}(t),$$

where $\widetilde{W}$ is a d-dimensional P-Wiener process. Thus (1.9) is an infinite dimensional stochastic system and, if we use the observed forward curve $\{f^*(0, T)\, ;\ T \geq 0\}$ as boundary value at $t = 0$, we will automatically have a perfect fit to data and the problem of inverting the yield curve is completely avoided. Moreover the following relations hold:

$$f(t, T) = -\frac{\partial \log p(t, T)}{\partial T}$$

or equivalently

$$p(t, T) = \exp\left\{-\int_t^T f(t, s)\, ds\right\}$$

and so specifying forward rates is equivalent to specifying bond prices.

It is easy to prove that, if we want that there exists a martingale measure Q for the bond market induced by the forward rate model (1.9), we have to impose some restrictions on the drift and the volatility. More precisely, if we give a model for the forward rate family directly under a martingale measure Q, that is, if we assume that the forward rates dynamics under Q is given by

$$(1.10) \qquad df(t, T) = \alpha(t, T)\, dt + \sigma(t, T)\, dW(t),$$

then the following condition, known as **"Heath - Jarrow - Morton drift condition"**, must be satisfied:

$$(1.11) \qquad \alpha(t, T) = \sigma(t, T) \int_t^T \sigma(t, s)\, ds,$$

for all $T > 0$ and all $t \leq T$.

Thus the volatility can be specified freely and then, the forward rate drift term is uniquely determined.

REMARK 1.5.1. The equality (1.11) should be interpreted as

$$\alpha(t, T) = \sum_{i=1}^{d} \sigma_i(t, T) \int_t^T \sigma_i(t, s) \, ds.$$

Finally we can observe that, if the forward rates dynamics under Q is given by (1.10), then the bond price dynamics under Q is given by

$$dp(t, T) = p(t, T) R(t) \, dt + p(t, T) S(t, T) \, dW(t),$$

where $S(t, T) = -\int_t^T \sigma(t, s) \, ds$.

Summing up, the problem of inverting the yield curve is avoided by working with forward rate models. However, there are other problems: the short rate will generally not be a Markov process and sometimes integration of (1.10) can be hard.

1.6. – Musiela parameterization

In the previous section the forward rates have been parameterized using the time *of* maturity. However, it is often much more useful to take the time *to* maturity, x, as parameter.

DEFINITION 1.6.1. *For all $x \geq 0$ the forward rates $r(t, x)$ are defined by the relation*

$$r(t, x) = f(t, t + x).$$

Now let us suppose that we have the following Q-dynamics for the forward rates:

$$df(t, T) = \left\{ \sigma(t, T) \int_t^T \sigma(t, s) \, ds \right\} dt + \sigma(t, T) \, dW(t).$$

Then (if f is sufficiently regular) the Q-dynamics of the family $r(t, x)$ is given by

$$dr(t, x) = \left\{ \frac{\partial}{\partial x} r(t, x) + \sigma_0(t, x) \int_0^x \sigma_0(t, y) \, dy \right\} dt + \sigma_0(t, x) \, dW(t),$$

where $\sigma_0(t, x) = \sigma(t, t + x)$.

Note that, when σ_0 is deterministic, this is a linear equation in an infinite dimensional space.

CHAPTER II

The Model

We take as given a filtered complete probability space $(\Omega, \mathcal{F}, Q, \underline{\mathcal{F}})$, where the filtration $\underline{\mathcal{F}}$ is the usual enlargement of the filtration generated by a d-dimensional Wiener process W.

In the sequel we are going to use the Musiela parameterization. Thus we denote with $p(t, x)$ the *price*, at time t, of a *zero coupon bond* maturing at $t + x$ and the *forward rate*, contracted at t, with maturity $t + x$, is defined by

$$r(t, x) = -\frac{\partial \log p(t, x)}{\partial x}.$$

Moreover we denote the *short rate* with R (i.e. we set $R(t) = r(t, 0)$) and we define the *yield* for the period $[t, t + x]$ as

$$y(t, x) = \int_0^x r(t, s) \, ds.$$

We consider a forward rate model given by the infinite dimensional stochastic differential equation

$$(2.1) \qquad dr(t, \cdot) = \alpha(t, \cdot) \, dt + \sigma(t, \cdot) \, dW(t), \qquad r(0, \cdot) = r^*(0, \cdot),$$

where $\alpha(t, \cdot)$ and $\sigma(t, \cdot)$ are such that the equation above has sense. The initial curve $\{r^*(0, x); \, x \geq 0\}$ is the observed forward rate curve.

We assume that the measure Q is a martingale measure for the bond market so that the model is free of arbitrage. Thus the Heath-Jarrow-Morton drift condition must hold and we have

$$\alpha(t, x) = \frac{\partial}{\partial x} r(t, x) + \sigma(t, x) \int_0^x \sigma(t, s)\, ds.$$

In the following chapters we will specify the model giving assumptions on the volatility σ.

CHAPTER III

Linear Realizations of Interest Rate Models

In this chapter, using results from systems theory, we give necessary and sufficient conditions so that the input-output map generated by a forward rate model with a *deterministic* volatility, has a *linear* finite dimensional realization. In other words, we give conditions under which the forward rate process $r(t, x)$ induced by a model $\mathcal{M}$ with a deterministic volatility, can be realized by a system of the form

$$\mathrm{d}z(t) = Az(t)\,\mathrm{d}t + B\,\mathrm{d}W(t)$$
$$r(t, x) = C(x)z(t)$$

where z is finite dimensional, A, B are matrices and $C(x)$ is a row-vector function. Note that the Wiener process W above is the same as in equation (2.1).

We give also a formula for the determination of the dimension of a minimal realization. Finally we illustrate an economic interpretation of the state space for a minimal realization in terms of a minimal set of benchmark forward rates.

3.1. – Assumptions

We consider the model defined in Chapter 2. In particular, we assume that the volatility $\sigma(t, x)$ is a deterministic time-invariant C^{∞}

function $\sigma(x)$ defined on $\mathbb{R}_+$ and with values in $\mathbb{R}^d$. Thus we have

(3.1)
$$dr(t) = \{Fr(t) + D\} \, dt + \sigma \, dW(t) \qquad r(0, x) = r^*(0, x)$$

$$y(t) = Hr(t)$$

where the linear operators F and H are defined by

$$F = \frac{\partial}{\partial x}, \qquad Hg(x) = \int_0^x g(s) \, ds$$

and D is a constant vector function in the state space $C[0, \infty)$, defined by $D(x) = \sigma(x) \int_0^x \sigma(s) \, ds = \frac{1}{2} F \| H\sigma \|^2$.

(Note that we have used the notation $r(t)$ (resp. $y(t)$) for $r(t, \cdot)$ (resp. $y(t, \cdot)$).)

Let us observe that system (3.1) is an infinite dimensional *linear* dynamical system which takes a Wiener trajectory $W(\cdot, \omega)$ (in $\mathbb{R}^d$) into a y-trajectory $y(\cdot, \star, \omega)$. So we have an input-output map Φ defined on $C[\mathbb{R}_+; \mathbb{R}^d]$ with value in $C[\mathbb{R}_+; \mathcal{H}]$, where $\mathcal{H} = C[0, \infty]$ is the space of yield curves.

The first problem is now to find when a *finite* dimensional stochastic differential equation induces the same input-output map as the one induced by system (3.1). At this end we need realization theory and so in the next section we sum up some results regarding this topic. For more information the reader is referred to any standard text on linear systems theory, such as e.g. [5].

3.2. – Systems theory and realization theory

Let us consider a finite dimensional system Σ:

$$\dot{z}(t) = Az(t) + Bu(t), \qquad z(0) = 0$$

$$y(t) = Cz(t)$$

where A is an $n \times n$-matrix, B is an $n \times d$-matrix and C is a $k \times n$-matrix.

The vector function u is the *input signal*, the vector function z is the *state vector* and y is the *output signal*.

DEFINITION 3.2.1. *The* **reachable subspace**, *$\mathcal{R}$, of the couple $[A, B]$ is the set of all points z in $\mathbb{R}^n$ which can be reached from the origin at any time using any input signal u.*

The system

$$\dot{z}(t) = Az(t) + Bu(t), \qquad z(0) = 0$$

or, more simply, the couple $[A, B]$ is termed **reachable** *if all points can be reached, that is, if $\mathcal{R} = \mathbb{R}^n$.*

Then we have the following propositions:

PROPOSITION 3.2.2. *The subspace $\mathcal{R}$ is given by*

$$\mathcal{R} = \text{Span} \left[B, \ AB, \ A^2 B, \ \ldots \right],$$

where the Span operation is interpreted as the linear hull of the column vectors of the matrices.

PROPOSITION 3.2.3. *The couple $[A, B]$ is reachable if and only if*

$$\text{Rank} \left[B, \ AB, \ A^2 B, \ \ldots, \ A^{n-1} B \right] = n.$$

We also need the following definition:

DEFINITION 3.2.4. *The* **silent subspace** *$\mathcal{S}$ of the system*

$$\dot{z}(t) = Az(t), \qquad z(0) = z_0$$

$$y(t) = Cz(t)$$

or, more simply, of the couple $[A, C]$ is defined by

$$\mathcal{S} = \{z_0 \in \mathbb{R}^n \mid y(t) \equiv 0\}.$$

We say that the couple $[A, C]$ is **observable** *if $\mathcal{S} = \{0\}$.*

Note that if $C : \mathbb{R}^n \mapsto \mathbb{R}^k$ is injective then $[A, C]$ is observable. More precisely, we have the following proposition:

PROPOSITION 3.2.5. *The couple* $[A, C]$ *is observable if and only if*

$$\text{Rank} \begin{bmatrix} C \\ CA \\ \vdots \\ CA^{n-1} \end{bmatrix} = n.$$

If we take the Laplace transforms in Σ, we obtain:

$$s\tilde{z} = A\tilde{z} + B\tilde{u}$$

$$\tilde{y} = C\tilde{z},$$

that is

$$\tilde{y} = C[sI - A]^{-1} B\tilde{u},$$

where $\tilde{z}$ (resp. $\tilde{u}$, $\tilde{y}$) denotes the Laplace transform of z (resp. u, y). The rational matrix function of degree ≤ -1 given by

$$G(s) = C[sI - A]^{-1} B$$

is termed the **transfer function** of Σ.

Moreover the **input-output map** Φ_Σ generated by the system Σ is the map defined by

$$\Phi_\Sigma : C[\mathbb{R}_+; \mathbb{R}^d] \longmapsto C[\mathbb{R}_+; \mathbb{R}^k]$$

$$[\Phi_\Sigma(u)](t) = y(t, u) = \int_0^t Ce^{A(t-s)} Bu(s) \, ds.$$

(In the same way, it is defined the input-output map generated by any other system like Σ.)

Now we are ready to give the definition of realization.

Let us consider an m-dimensional system Γ:

$$\dot{r}(t) = Fr(t) + Gu(t), \qquad r(0) = 0$$

$$y(t) = Hr(t).$$

Then we have the following definition:

DEFINITION 3.2.6. *The n-dimensional system Σ given by*

$$\dot{z}(t) = Az(t) + Bu(t), \qquad z(0) = 0$$

$$y(t) = Cz(t)$$

is a **realization** *of the system Γ if the input-output map, Φ_Γ, of Γ coincides with the input-output map, Φ_Σ, of Σ. Moreover, we say that Σ is a* **minimal** *realization of Γ if the state space of Σ is of minimal dimension. The* **MacMillan degree** *of the system Γ is the dimension of a minimal realization.*

We have the following proposition:

PROPOSITION 3.2.7. *A realization, Σ, of the system Γ is minimal if and only if it is reachable and observable.*

Finally we can observe that Σ is a realization of Γ if and only if the two tranfer functions G_Σ and G_Γ coincide. Thus, in order to find a minimal realization of Σ, it is enough to find a triplet $\{A, B, C\}$ with A of minimal dimension such that $G(s) = C[sI - A]^{-1}B$. With regard to this, we recall the following proposition:

PROPOSITION 3.2.8. *For every rational matrix function $G(s)$ of degree ≤ -1 there exists a system Σ, or equivalently a triplet $\{A, B, C\}$, such that $G = G_\Sigma$, that is*

$$G(s) = C[sI - A]^{-1}B,$$

$$\left[\mathcal{L}^{-1}(G)\right](t) = Ce^{At}B$$

(where $\mathcal{L}$ is the Laplace transform.)

3.3. – Realizations of interest rate models

We now come back to our original problem and we consider the system (3.1).

Let us recall that a stochastic differential equation of the form

$$dy(t) = \{ay(t) + b\}\, dt + c\, dW(t)$$

has the solution

$$y(t) = e^{at} y(0) + \int_0^t e^{a(t-s)} b \, ds + \int_0^t e^{a(t-s)} c \, dW(s).$$

Thus we are led to say that the formal solution of equation (3.1) is the following:

$$(3.2) \quad r(t, x) = e^{Ft} r^*(0, x) + \int_0^t e^{F(t-s)} D(x) \, ds + \int_0^t e^{F(t-s)} \sigma(x) dW(s).$$

Now we have to say how the operator e^{Ft} acts on real valued functions f in $C[0, \infty)$. From the standard series expansion of the exponential function, we are led to write

$$\left[e^{Ft} f \right](x) = \sum_{n=0}^{\infty} \frac{t^n}{n!} \left[F^n f \right](x),$$

and so, if f is analytic, we have

$$\left[e^{Ft} f \right](x) = \sum_{n=0}^{\infty} \frac{t^n}{n!} \frac{\partial^n f}{\partial x^n}(x) = f(t + x).$$

Thus we find that

$$(3.3) \quad r(t, x) = r^*(0, x+t) + \int_0^t D(x+t-s) \, ds + \int_0^t e^{F(t-s)} \sigma(x) \, dW(s).$$

This is a heuristic proof of the following precise result (which can be proved rigorously).

PROPOSITION 3.3.1. *The operator F is the infinitesimal generator of the semigroup of left translation, that is, for any function f in C*[0, ∞), *we have*

$$\left[e^{Ft} f \right](x) = f(t + x).$$

Moreover the solution of equation (3.1) is given by (3.2) or, equivalently, by (3.3).

We can write (3.3) as

$$r(t, x) = r_0(t, x) + \delta(t, x),$$

where

$$\delta(t, x) = r^*(0, x + t) + \int_0^t D(x + t - s)\, ds$$

and

$$dr_0(t) = Fr_0(t, x)\, dt + \sigma(x)\, dW(t), \qquad r_0(0, x) = 0.$$

Moreover, we can write

$$y(t, x) = y_0(t, x) + \Delta(t, x),$$

where

$$\Delta(t, x) = \int_0^x \delta(t, u)\, du$$

and

$$y_0(t, x) = Hr_0(t, x).$$

Since we do not have the input W in $\Delta(t, x)$, in order to study the input-output behaviour of the system (3.1), it is enough to consider the system

(3.4)
$$dr_0(t, x) = Fr_0(t, x)\, dt + \sigma(x)\, dW(t), \quad r_0(0, x) = 0$$

$$y_0(t, x) = Hr_0(t, x).$$

DEFINITION 3.3.2. *A triplet* $\{A, B, C(x)\}$, *where A is an $n \times n$-matrix, B is an $n \times d$-matrix and $C(x)$ an n-dimensional row-vector function, is called an n-dimensional* **realization of the (r_0, y_0)-system** *if y_0 has the representation*

(3.5)
$$dz(t) = Az(t)\, dt + B\, dW(t), \quad z(0) = 0$$

$$y_0(t, x) = C(x)z(t).$$

Note that in the system (3.5), the infinite dimensionality is all in the output.

DEFINITION 3.3.3. *The triplet* $\{A, B, C(x)\}$ *is called an n-dimensional* **realization of the r_0-system** *given by*

$$dr_0(t, x) = Fr_0(t, x)\, dt + \sigma(x)\, dW(t), \qquad r_0(0, x) = 0,$$

if r_0 has the representation

$$dz(t) = Az(t)\, dt + B\, dW(t), \quad z(0) = 0$$

$$r_0(t, x) = C(x)z(t).$$

REMARK 3.3.4. Obviously $\{A, B, C(x)\}$ is a finite dimensional realization for r_0 if and only if $\{A, B, HC(x)\}$ is a realization for the (r_0, y_0)-system. Thus it is enough to consider only realizations of the r_0-system.

3.4. – Existence of finite dimensional realizations

In this section, we will study the existence of a finite dimensional realization for the r_0-system.

The following result links a realization of the r_0-system with a realization of a deterministic system.

LEMMA 3.4.1. *The system*

$$dz(t) = Az(t)\, dt + B\, dW(t), \quad z(0) = 0$$

(3.6)

$$r_0(t, x) = C(x)z(t).$$

is a realization of

(3.7) $$dr_0(t, x) = Fr_0(t, x)\, dt + \sigma(x)\, dW(t), \qquad r_0(0, x) = 0,$$

if and only if the deterministic system

(3.8) $$\frac{dr_0}{dt}(t, x) = Fr_0(t, x)\, dt + \sigma(x)u(t), \qquad r_0(0, x) = 0,$$

has the same input-output map as

(3.9)
$$\frac{dz}{dt}(t) = Az(t) + Bu(t), \qquad z(0) = 0$$

$$r_0(t, x) = C(x)z(t).$$

PROOF. We have $r_0(t, x) = \int_0^t \sigma(x + t - s) \, dW(s)$. Then, integrating by parts, we find

$$r_0(t, x) = W(t)\sigma(x) + \int_0^t W(s)\sigma'(x + t - s) \, ds.$$

It follows that $r_0(t, x, \omega) = \Phi(x, W(\cdot, \omega))(t)$, where the mapping $\Phi(x, \cdot)$: $C[0, \infty) \mapsto C[0, \infty)$, is defined by

$$\Phi(x, v)(t) = v(t)\sigma(x) + \int_0^t v(s)\sigma'(x + t - s) \, ds.$$

We now observe that, for each x, the mapping $\Phi(x, \cdot)$ is continuous in the topology of uniform convergence on compacts and, so, it is completely determined by its behaviour on $C^1[0, \infty)$ (since it is a dense subset in $C[0, \infty)$). On the other hand, for any v in $C^1[0, \infty)$, we have $\Phi(x, v)(t) = r_0(t, x)$, where

$$\frac{dr_0}{dt}(t, x) = Fr_0(t, x) \, dt + \sigma(x)u(t), \qquad r_0(0, x) = 0,$$

and $u(t) = \frac{dv}{dt}(t)$. So we can conclude. □

Because of the previous lemma, we can now translate the problem in terms of transfer functions. To this end, let us consider the deterministic system (3.8). The **transfer function** of the system (3.8) is the function $G(s, x)$ defined by the relation

$$\widetilde{r}_0(s, x) = G(s, x)\widetilde{u}(s),$$

where $\widetilde{r}_0$ (resp. $\widetilde{u}$) denotes the Laplace transform of r_0 (resp. u).

Moreover the system (3.9) has the same input-output map as the system (3.8) if and only if they have the same transfer function. Thus we have the following corollary:

COROLLARY 3.4.2. *The system* (3.6) *is a realization of* (3.7) *if and only if the system* (3.9) *has the same transfer function as* (3.8).

Let us now determine the transfer function for the system (3.8).

PROPOSITION 3.4.3. *The transfer function* $G(s, x)$ *of the system* (3.8) *is given by*

$$G(s, x) = \mathcal{L} \{\sigma_x\} (s),$$

where $\mathcal{L}$ *denotes the Laplace transform and the subscript denotes translation, that is* $g_x(t) = g(t + x)$.

PROOF. We have

$$r_0(t, x) = \int_0^t \sigma(x + t - s)u(s)\, ds = [\sigma_x * u](t)$$

and thus we get $\widetilde{r}_0(s, x) = \mathcal{L}\{\sigma_x\}(s)\,\widetilde{u}(s).$ □

We may now give the following main result:

PROPOSITION 3.4.4. *The system*

$$dr_0(t, x) = Fr_0(t, x)\, dt + \sigma(x)\, dW(t), \qquad r_0(0, x) = 0$$

has a finite dimensional realization if and only if the volatility function σ *can be written in the form*

$$\sigma(x) = Ce^{Ax}B,$$

where A, B *are matrices and* C *is a row-vector.*
 Moreover, if σ *has the form above, then a concrete realization is given by*

$$dz(t) = Az(t)\, dt + B\, dW(t), \quad z(0) = 0$$

$$r_0(t, x) = C(x)z(t)$$

with A, B *as above and with* $C(x) = Ce^{Ax}$.

PROOF. Let us recall that the transfer function of (3.9) is given by

$$H(s, x) = C(x)\, [sI - A]^{-1}\, B.$$

Thus, if (3.6) is a realization of (3.7), then we must have

$$G(s, x) = H(s, x) = C(x)\, [sI - A]^{-1}\, B.$$

On the other hand, because of Proposition 3.4.3, we have

$$G(s, x) = \mathcal{L}\{\sigma_x\}(s)$$

and so, we may conclude that $\sigma_x(t)$ is of the form $C(x)e^{At}B$. It follows that

$$\sigma(t) = \sigma_0(t) = C(0)e^{At}B = Ce^{At}B.$$

Now let us suppose that σ has the form $\sigma(x) = Ce^{Ax}B$. Then we have $\sigma_x(t) = Ce^{Ax}e^{At}B$ and so we have

$$G(s, x) = \mathcal{L}\{\sigma_x\}(s) = Ce^{Ax}[sI - A]^{-1}B.$$

But this is the transfer function of the system (3.9) with $C(x) = Ce^{Ax}$. □

3.5. – Minimal realizations

In this section we want to determine the minimal dimension of a finite dimensional realization of the r_0-system (3.7).

Let us start giving the following definition:

DEFINITION 3.5.1. *The* **dimension** *of a realization* $\{A, B, C(x)\}$ *(of the r_0-system) is defined as the dimension of the corresponding state space. A realization* $\{A, B, C(x)\}$ *is said to be* **minimal** *if there is no other realization with smaller dimension. The* **MacMillan degree**, μ, *of the r_0-system is defined as the dimension of a minimal realization.*

We have the following result:

PROPOSITION 3.5.2. *The MacMillan degree,* μ, *is given by*

$$\mu = \dim \mathcal{R},$$

where

$$\mathcal{R} = \mathrm{Span}\left[F^k \sigma_i \; ; \; i = 1, \ldots, d \quad k \geq 0 \right]$$

with $F = \frac{\mathrm{d}}{\mathrm{d}x}$.

PROOF. It is obvious that $\mathcal{R}$ has a finite dimension if and only if σ satisfies a linear ordinary differential equation with constant coefficients and so, if and only if σ is of the form $\sigma(x) = Ce^{Ax}B$. It follows that $\mathcal{R}$ has a finite dimension if and only if there exists a finite dimensional realization.

Let us now assume that $\mu = n$ and that $\{A, B, Ce^{Ax}\}$ is a minimal finite dimensional realization. Then (because of Proposition 3.2.7)

[A, B] is reachable and [A, C] is observable. Moreover, by Proposition 3.4.4, we can write $\sigma(x) = Ce^{Ax}B$ and so, we have

$$ F^k\sigma(x) = Ce^{Ax}A^kB, \qquad k = 0, 1, \ldots . $$

Let us now consider the linear mapping $\Lambda : \mathbb{R}^n \mapsto C[0, \infty)$ defined by

$$ [\Lambda b] = Ce^{Ax}b. $$

We can observe that, $\Lambda b = 0$ (i.e. $Ce^{Ax}b = 0$ for all x), then, taking derivatives and setting $x = 0$, we find the equation

$$ \begin{bmatrix} C \\ CA \\ \vdots \\ CA^{n-1} \end{bmatrix} b = 0, $$

and so $b = 0$ since [A, C] is observable. Then the map Λ is injective and so

$$ \dim \mathcal{R} = \dim \mathrm{Span} \left[F^k\sigma_i \; ; \; i = 1, \ldots, d \quad k \geq 0 \right] $$

$$ = \dim \Lambda \left(\mathrm{Span} \left[B, \; AB, \; A^2B, \ldots \right] \right) $$

$$ = \dim \mathrm{Span} \left[B, \; AB, \; A^2B, \ldots \right]. $$

On the other hand, [A, B] is reachable and so

$$ \mathrm{Span}[B, \; AB, \; A^2B, \ldots] = \mathbb{R}^n, $$

that is $\dim \mathrm{Span} \left[B, \; AB, \; A^2B, \ldots \right] = n$. $\qquad\qquad\qquad\square$

EXAMPLE 3.5.3. **(Hull-White extended Vasiček model)**
If we have $\sigma(x) = \sigma e^{-ax}$ (i.e. $A = -a$, $B = \sigma$, $C = 1$), then

$$\mathcal{R} = \text{Span}\left[\sigma e^{-ax}, -\sigma a e^{-ax}, \sigma a^2 e^{-ax}, \ldots\right]$$

and so we find $\mu = 1$. Moreover, by Proposition 3.4.4, we find the following realization:

$$dz(t) = -az(t)\,dt + \sigma\,dW(t), \quad z(0) = 0$$

$$r_0(t, x) = e^{-ax}z(t).$$

We can observe that, since the state space in this realization is of dimension one, it is minimal.

3.6. – Economic interpretation of the state space

The state space of a minimal realization of the r_0-system has a very simple economic interpretation. More precisely, we have the following results.

PROPOSITION 3.6.1. *Assume that $\{A, B, Ce^{Ax}\}$ is a minimal realization of the r_0-system, with dimension n. Then it is possible to choose arbitrarily a set of distinct "benchmark" times of maturity $x_1, \ldots, x_n$ such that*

$$T(\bar{x}) = \begin{bmatrix} Ce^{Ax_1} \\ \vdots \\ Ce^{Ax_n} \end{bmatrix}$$

is invertible.

PROPOSITION 3.6.2. *Let us use the notation*

$$r_0(t, \bar{x}) = \begin{bmatrix} r_0(t, x_1) \\ \vdots \\ r_0(t, x_n) \end{bmatrix}$$

and corresponding interpretations for the vector $r(t, \bar{x})$ and $\delta(t, \bar{x})$.

Then the vector $r(t, \bar{x})$ of benchmark forward rates has the dynamics

$$dr(t, \bar{x}) = \left\{ T(\bar{x}) A T^{-1}(\bar{x}) r(t, \bar{x}) + \psi(t, \bar{x}) \right\} dt + T(\bar{x}) B \, dW(t)$$

$$r(0, \bar{x}) = r^*(0, \bar{x}),$$

where the deterministic function $\psi(t, x)$ is given by

$$\psi(t, \bar{x}) = \frac{\partial}{\partial t} \delta(t, \bar{x}) - T(\bar{x}) A T^{-1}(\bar{x}) \delta(t, \bar{x}).$$

(The function $\delta(t, x)$ is defined in Section 3.3.)

Moreover the system of benchmark forward rates gives the entire forward rate process according to the formula

$$r(t, x) = C e^{Ax} T^{-1}(\bar{x}) r(t, \bar{x}) - C e^{Ax} T^{-1}(\bar{x}) \delta(t, \bar{x}) + \delta(t, x).$$

Finally we have an affine term structure:

$$y(t, x) = \alpha(x) r(t, \bar{x}) - \beta(x).$$

For the proof of these results see [3].

EXAMPLE 3.6.3.   Let us consider again the case $\sigma(x) = \sigma e^{-ax}$. We have $\mu = 1$ (see Example 3.5.3) and so we can choose a single benchmark maturity, for example we can take $x_1 = 0$ (i.e. we take the short rate $R(t) = r(t, 0)$ as the state variable). We then have $T(\bar{x}) = 1$ and, by Proposition 3.6.2, we find

$$dR(t) = \{-aR(t) + \psi(t, 0)\} \, dt + \sigma \, dW(t).$$

Thus we see that we indeed have the Hull-White extended Vasiček model.

CHAPTER IV

Interest Rate Dynamics and Consistent Forward Rate Curves

In this chapter we work in a context like the one described in Chapter 2, that is we consider an arbitrage free forward rate model $\mathcal{M}$ given by

$$dr(t, x) = \left\{ \frac{\partial}{\partial x} r(t, x) + \sigma(t, x) \int_0^x \sigma(t, s)\, ds \right\} dt + \sigma(t, x)\, dW(t)$$

$$r(0, x) = r^*(0, x),$$

where x denotes the time to maturity and $\{r^*(0, x)\,;\, x \geq 0\}$ is the observed forward rate curve at $t = 0$.

A standard procedure when dealing with concrete interest rate models on a high frequency (say, daily) basis can be described as follows:

1. At time $t = 0$, use market data to fit (calibrate) the model to the observed bond prices.

2. Use the calibrated model to compute prices of various interest rate derivatives.

3. The following day $(t = 1)$, repeat the procedure in 1. above in order to recalibrate the model, etc.

To carry out the calibration in Step 1. above, the analyst typically has to produce a forward rate curve $\{r^*(0, x)\,;\, x \geq 0\}$ from the observed

data. However, since only a finite number of bonds is actually traded
in the market, the data consist of a discrete set of points, and a need to
fit a curve to these points arises. This curve-fitting may be done in a
variety of ways. One way is to use splines, but also a number of parameterized families of smooth forward rate curves have become popular in
applications — the most well-known probably being the Nelson-Siegel
(see [12]) family. Once the curve $\{r^*(0, x); x \geq 0\}$ has been obtained,
the parameters of the interest rate model may be calibrated to this.

Now, from a purely logical point of view, the recalibration procedure in Step 3 above is of course slightly nonsensical: If the interest
rate model at hand is an exact picture of reality, then there should be
no need to recalibrate. The reason that everyone insists on recalibrat-
ing is of course that any model in fact only is an approximate picture
of the financial market under consideration, and recalibration allows in-
corporating newly arrived information in the approximation. Even so,
the calibration procedure itself ought to take into account that it will
be repeated. It appears that the optimal way to do so would involve
a combination of time series and cross-section data, as opposed to the
purely cross-sectional curve-fitting, where the information contained in
previous curves is discarded in each recalibration.

The cross-sectional fitting of a forward curve and the repeated recalibration is thus, in a sense, a pragmatic and somewhat non-theoretical
endeavour. Nonetheless, there are some nontrivial theoretical problems
to be dealt with in this context, and the problem to be studied in this
chapter concerns the *consistency* between, on the one hand, the dynamics
of a given interest rate model, and, on the other hand, the forward curve
family employed.

What, then, is meant by consistency in this context? Assume that
a given interest rate model $\mathcal{M}$ (e.g. the Hull-White model) in fact *is* an
exact picture of the financial market. Now consider a particular family
$\mathcal{G}$ of forward rate curves (e.g. the Nelson-Siegel family) and assume
that the interest rate model is calibrated using this family. We then say
that the pair $(\mathcal{M}, G)$ is **consistent** (or, that $\mathcal{M}$ and $\mathcal{G}$ are consistent) if
all forward curves which may be produced by the interest rate model
$\mathcal{M}$ are contained within the family $\mathcal{G}$. Otherwise, the pair $(\mathcal{M}, \mathcal{G})$ is
inconsistent.

Thus, if $\mathcal{M}$ and $\mathcal{G}$ are consistent, then the interest rate model actually
produces forward curves which belong to the relevant family. In contrast,
if $\mathcal{M}$ and $\mathcal{G}$ are inconsistent, then the interest rate model will produce
forward curves outside the family used in the calibration step, and this
will force the analyst to change the model parameters all the time —

not because the model is an approximation to reality, but simply because the family does not go well with the model.

Put into more operational terms this can be rephrased as follows.

- Suppose that you are using a fixed interest rate model $\mathcal{M}$. If you want to do recalibration, then your family $\mathcal{G}$ of forward rate curves should be chosen in such a way as to be consistent with the model $\mathcal{M}$.

Note however that the argument also can be run backwards, yielding the following conclusion for empirical work.

- Suppose that a particular forward curve family $\mathcal{G}$ has been observed to provide a good fit, on a day-to-day basis, in a particular bond market. Then this gives you modelling information about the choice of an interest rate model in the sense that you should try to use/construct an interest rate model which is consistent with the family $\mathcal{G}$.

We now have a number of natural problems to study.

I Given an interest rate model $\mathcal{M}$ and a family of forward curves $\mathcal{G}$, what are necessary and sufficient conditions for consistency?

II Take as given a specific family $\mathcal{G}$ of forward curves (e.g. the Nelson-Siegel family). Does there exist any interest rate model $\mathcal{M}$ which is consistent with $\mathcal{G}$?

III Take as given a specific interest rate model $\mathcal{M}$ (e.g. the Hull-White model). Does there exist any finitely parameterized family of forward curves $\mathcal{G}$ which is consistent with $\mathcal{M}$?

In this chapter we are concerned with the first two questions, while in Chapter 5 we consider the last question.

4.1. – The formal problem statement

We denote the forward rate curve at time t by r_t and we consider the special case when forward rate dynamics is of the form

$$(4.1) \qquad dr_t = \alpha(r_t)\, dt + \sigma(r_t)\, dW_t,$$

where

$$\alpha(r) = \alpha(r, x) = \frac{\partial}{\partial x} r + \sigma(r, x) \int_0^x \sigma(r, s) \, ds$$

and

$$\sigma(r) = \sigma(r, x).$$

It is an infinite dimensional stochastic differential equation evolving in the space $\mathcal{H}$ of forward rate curves and specifying the model (4.1) is equivalent to specifying the volatility function $\sigma(r, x) : \mathcal{H} \times \mathbb{R}_+ \mapsto \mathbb{R}^d$.

We also consider as given a smooth mapping

$$G : \mathbb{R}^m \longmapsto \mathcal{H}$$
$$z \longmapsto G(\star \, ; z),$$

that is a finitely parameterized family of forward rate curves.

We take as the space $\mathcal{H}$ one of the spaces $\mathcal{H}_\gamma$ defined as follows:

DEFINITION 4.1.1. *Consider a fixed real number $\gamma > 0$. The space $\mathcal{H}_\gamma$ is defined as the space of all differentiable (in the distributional sense) functions $r : \mathbb{R}_+ \longmapsto \mathbb{R}$ such that*

$$\|r\|_\gamma^2 = \int_0^\infty r^2(x) \, e^{-\gamma x} \, dx + \int_0^\infty \left(\frac{dr}{dx}(x) \right)^2 e^{-\gamma x} \, dx < \infty.$$

REMARK 4.1.2. The space $\mathcal{H}_\gamma$ with the inner product

$$(q, r) = \int_0^\infty q(x) r(x) \, e^{-\gamma x} \, dx + \int_0^\infty \left(\frac{dq}{dx}(x) \right) \left(\frac{dr}{dx}(x) \right) e^{-\gamma x} \, dx$$

is a Hilbert space. Moreover, because of the exponential weighting function, all smooth and bounded functions belong to $\mathcal{H}_\gamma$.

REMARK 4.1.3. We have taken, for simplicity, a mapping G defined on the entire space $\mathbb{R}^m$. In reality, the results illustrated below hold also for maps $G : \mathcal{Z} \longmapsto \mathcal{H}$, where $\mathcal{Z}$ is an open connected subset of $\mathbb{R}^m$. Moreover we have assumed that σ is time-invariant but the case when σ is of the form $\sigma(t, r, x)$ can be treated in exactly the same way (see [2]).

EXAMPLE 4.1.4. **(Nelson-Siegel)** The Nelson-Siegel forward curve family is given by

$$G(x, z) = z_1 + z_2 e^{-z_4 x} + z_3 x e^{-z_4 x}, \qquad z \in \mathbb{R}^4.$$

But, if we want the image of this map to be included in $\mathcal{H}_\gamma$, then we need to impose the restriction $z_4 > -\gamma/2$. Moreover, as we shall see below, we also need to impose $z_4 \neq 0$ so that G and its Fréchet derivative G'_z are injective. Thus we take as the parameter space for the Nelson-Siegel family the space $\mathcal{Z} = \{z \in \mathbb{R}^4 : z_4 \neq 0,\ z_4 > -\frac{\gamma}{2}\}$.

DEFINITION 4.1.5. *Given the mapping $G : \mathbb{R}^m \longmapsto \mathcal{H}$, the* **forward curve manifold** $\mathcal{G} \subseteq \mathcal{H}$ *is defined as*

$$\mathcal{G} = \mathrm{Im}[G] = \{G(\cdot\,;z) \in \mathcal{H}\,;\ z \in R^m\}.$$

Thus $\mathcal{G}$ is the set of all forward rate curves produced by G.

DEFINITION 4.1.6. *We say that $\mathcal{G}$ is* **locally invariant** *under the action of r if, for each point $(s, r) \in \mathbb{R}_+ \times \mathcal{G}$, there exists a stopping time $\tau(s, r)$ such that $\tau(s, r) > s$, Q-a.s. and the following implication holds Q-a.s.:*

$$r_s \in \mathcal{G} \Longrightarrow r_t \in \mathcal{G} \quad \text{for all } t \text{ with } s \leq t < \tau(s, r_s).$$

If $\tau(s, r) = +\infty$ for all $(s, r) \in \mathbb{R}_+ \times \mathcal{G}$, Q-a.s., we say that $\mathcal{G}$ is (globally) **invariant**.

Our main problem is now to investigate when the forward rate curve manifold $\mathcal{G}$, given by a mapping G, is invariant under the action of r.

In the sequel we need the following assumptions:

ASSUMPTIONS. *Let G'_x and G'_z denote the Fréchet derivatives of G with respect to x and z, respectively. We assume the following:*

1. *The map $r \mapsto \sigma(r)$ is a C^3 map from $\mathcal{H}$ to $\mathcal{H}^d$.*

2. *For every initial point r_0 in $\mathcal{G}$, there exists a unique strong solution in $\mathcal{H}$ of equation (4.1).*

3. *The map $z \mapsto G(z)$ is injective and the Fréchet derivative $G'_z(z)$ is injective for all z in $\mathbb{R}^m$.*

4. *The map $z \mapsto G'_x(z)$ is a continuous map from $\mathbb{R}^m$ to $\mathcal{H}$.*

These hypotheses are assumed throughout this chapter.

4.2. – The deterministic finite dimensional case

In order to get some intuition, let us start with a sketch of what it happens in the deterministic finite dimensional case.

Let us consider an n-dimensional process satisfying an ordinary differential equation of the form

$$\frac{dr}{dt} = \alpha(t, r).$$

Let us take as given also a smooth mapping $G : \mathbb{R}^m \mapsto \mathcal{H} = \mathbb{R}^n$, with $m < n$. As before, we define $\mathcal{G}$ as the m-dimensional surface in $\mathcal{H}$ given by $\mathcal{G} = \{G(z) : z \in \mathbb{R}^m\}$. Then we have the following obvious proposition:

PROPOSITION 4.2.1. *The manifold $\mathcal{G}$ is invariant under the action of r (i.e. we have $r(0) \in \mathcal{G} \implies r(t) \in \mathcal{G}$ for all t) if and only if the velocity vector $\frac{dr}{dt}$ belongs to the tangent space $T_{\mathcal{G}}(r(t))$ for each $t > 0$.*

Now we can observe that a generic point of the manifold $\mathcal{G}$ is written as $y = G(z)$ and the tangent space at this point is given as the linear hull of the tangent vectors

$$\frac{\partial G(z)}{\partial z_i}, \qquad i = 1, \ldots, m.$$

Then the tangent space $T_{\mathcal{G}}(G(z))$ coincides with $\mathrm{Im}\left[G'_z(z)\right]$, where $G'_z(z)$ is the Jacobian of G at z, and so we have the following proposition:

PROPOSITION 4.2.2. *The manifold $\mathcal{G}$ is invariant under the action of r if and only if, for each z in $\mathbb{R}^m$ and each $t > 0$, we have*

$$\alpha(t, G(z)) \in \mathrm{Im}\left[G'_z(z)\right].$$

Analogously, if we consider the n-dimensional control system

$$\frac{dr}{dt} = \alpha(t, r) + \sigma(t, r)u,$$

where the input signal u can be chosen arbitrarily, we have the following two propositions:

PROPOSITION 4.2.3. *The manifold $\mathcal{G}$ is invariant under the action of r, for all input signals u, if and only if the velocity vector of r is in the tangent space of $\mathcal{G}$ for each choice of the input signal u.*

PROPOSITION 4.2.4. *The manifold $\mathcal{G}$ is invariant under the action of r if and only if, for each z in $\mathbb{R}^m$ and each $t > 0$, we have*

$$\alpha\,(t, G(z)) + \sigma\,(t, G(z))\,u \in \operatorname{Im}\left[G_z'(z)\right], \qquad u \in \mathbb{R}^d$$

or, equivalently,

$$\alpha\,(t, G(z)) \in \operatorname{Im}\left[G_z'(z)\right]$$
$$\sigma\,(t, G(z)) \in \operatorname{Im}\left[G_z'(z)\right]$$

(where the last relation is interpreted as holding columnwise, that is, $\sigma_j\,(t, G(z)) \in \operatorname{Im}\left[G_z'(z)\right]$ for each $j \in \{1, \ldots, d\}$).

4.3. – The stochastic finite dimensional case

Let us now consider an n-dimensional process r satisfying a stochastic differential equation of the form

$$(4.2) \qquad\qquad dr_t = \alpha(r_t)\,dt + \sigma(r_t)\,dW_t,$$

with W a d-dimensional Wiener process and α, σ smooth vector fields. We also take as given a smooth mapping $G : \mathbb{R}^m \longmapsto \mathbb{R}^n$ (with $m < n$).

Arguing informally, we can write (4.2) as

$$\frac{dr}{dt} = \alpha(t, r) + \sigma(t, r)u,$$

where $u = \frac{dW}{dt}$ is interpreted as a randomly chosen input. Because of what we have seen in Section 4.2, we are led to the following conjecture:

The manifold $\mathcal{G}$ is invariant under the action of r if and only if, for each z in $\mathbb{R}^m$ and $t > 0$, we have the relations

$$\alpha\,(t, G(z)) \in \operatorname{Im}\left[G_z'(z)\right]$$
$$\sigma\,(t, G(z)) \in \operatorname{Im}\left[G_z'(z)\right].$$

However, this conjecture is **wrong**, since it neglects the difference between ordinary differential calculus and Itô calculus. Therefore, we need to rewrite everything in terms of Stratonovich integrals.

DEFINITION 4.3.1. *For given continuous semimartingales X and Y, the* **Stratonovich integral** *of X with respect to Y, denoted by* $\int_0^t X(s) \circ dY(s)$, *is defined as*

$$\int_0^t X(s) \circ dY(s) = \int_0^t X(s)\,dY(s) + [X, Y]_t.$$

Since in our case we have only Wiener processes as driving noise, we can define the quadratic variation process $[X, Y]$ by

$$d[X, Y]_t = dX(t)dY(t),$$

with the usual "multiplication rules": $dW dt = dt dt = 0$, $dW dW = dt$.

For the Stratonovich integral, we have the following "chain rule":

PROPOSITION 4.3.2. *For any smooth function $F(t, y)$ we have*

$$dF(t, Y_t) = \frac{\partial F}{\partial t}(t, Y_t)\,dt + \frac{\partial F}{\partial y} \circ dY_t.$$

Using the Stratonovich integral, we find the following result:

PROPOSITION 4.3.3. *Assume that the process r has a Stratonovich differential given by*

$$dr = \alpha(t, r)\,dt + \sigma(t, r) \circ dW(t).$$

Then the manifold $\mathcal{G}$ is invariant for r if and only if, for each z in $\mathbb{R}^m$ and $t > 0$, we have the following relations

$$\alpha(t, G(z)) \in \text{Im}\left[G_z'(z)\right]$$
$$\sigma(t, G(z)) \in \text{Im}\left[G_z'(z)\right].$$

This result is a special case of the result for infinite dimensional case illustrated below.

4.4. – Invariant forward curve manifold

We now return to our original infinite dimensional problem.

Given the Itô dynamics (4.1), the Stratonovich dynamics is given by

(4.3)
$$dr_t(x) = \left\{ \frac{\partial}{\partial x} r_t(x) + \sigma(r_t, x) \int_0^x \sigma(r_t, y) dy - \frac{1}{2} [\sigma'_r(r_t)\sigma(r_t)](x) \right\} dt$$

$$+ \sigma(r_t, x) \circ dW(t).$$

REMARK 4.4.1. In order to get equation (4.3), we have used the infinite dimensional Itô formula (see [6]) to find the Itô dynamics of σ,

$$d\sigma(r_t) = \{\ldots\} dt + \sigma'_r(r_t)\sigma(r_t) dW(t),$$

where $\sigma'_r(r)\sigma(r)$ denotes the Fréchet derivative $\sigma'_r(r)$ of σ (with respect to the infinite dimensional r-variable) operating on $\sigma(r)$.

We note also that in the deterministic volatility case, $\sigma = \sigma(x)$, we have $\sigma'_r = 0$ and Itô and Stratonovich dynamics coincide.

Let us give the following version of invariance:

DEFINITION 4.4.2. *Consider a fixed manifold* $\mathcal{G} = \text{Im}[G]$ *(where* G : $\mathbb{R}^m \mapsto \mathcal{H}$ *satisfies the assumptions stated in 4.1) and a fixed r-model of the form*
$$dr = \alpha(t, r) dt + \sigma(t, r) \circ dW(t).$$

We say that $\mathcal{G}$ *is* **locally r-invariant** *under the action of r if, for each initial curve* r_0 *in* $\mathcal{G}$, *there exists a strictly positive (Q-a.s.) stopping time* $\tau(r_0)$, *and a stochastic process Z on* $\mathbb{R}^m$ *with a Stratonovich differential of the form*

$$dZ_t = a(Z_t) dt + b(Z_t) \circ dW_t$$

such that we have (Q-a.s.) the representation

$$r_t(x) = G(x; Z_t), \qquad \forall x, \qquad \forall 0 \leq t < \tau(r_0).$$

If $\tau(r) = +\infty$ *for all r (Q-a.s.), we say that* $\mathcal{G}$ *is (globally)* **r-invariant**.

It is obvious that local r-invariance implies local invariance, and in fact they can be shown to be equivalent (see [2]).

Now we can state the main result.

THEOREM 4.4.3. *The forward curve manifold $\mathcal{G}$ is locally invariant under r if and only if the conditions*

$$(4.4) \qquad G_x'(z) + D(r) - \frac{1}{2}\sigma_r'(r)\sigma(r) \in \mathrm{Im}\left[G_z'(z)\right],$$

$$(4.5) \qquad\qquad\qquad \sigma(r) \in \mathrm{Im}\left[G_z'(z)\right],$$

with $D(r, x) = \sigma(r_t, x) \int_0^x \sigma(r_t, y)\,\mathrm{d}y$ and $r = G(z)$, hold for all z in $\mathbb{R}^m$.

Condition (4.5) is interpreted as holding componentwise. Condition (4.4) is called **the consistent drift condition** and (4.5) is called **the consistent volatility condition**.

The proof of the result above can be found in [2].

4.5. – Concrete examples

In this section, we apply the previous result to a specific forward curve family, the Nelson-Siegel family, and to two given interest rate models, Hull-White extended Vasiček and Ho-Lee interest rate models.

The *Nelson-Siegel* (NS) forward curve family is given by

$$(4.6) \qquad G(x, z) = z_1 + z_2 e^{-z_4 x} + z_3 x e^{-z_4 x}, \qquad z \in \mathcal{Z}$$

where $\mathcal{Z}$ is the set $\{z \in \mathbb{R}^4 : z_4 \neq 0, z_4 > -\frac{\gamma}{2}\}$ (with $\gamma > 0$).
The Fréchet derivatives are the following:

$$(4.7) \qquad G_z'(z, x) = \left[1, e^{-z_4 x}, x e^{-z_4 x}, -(z_2 + z_3 x)x e^{-z_4 x}\right],$$

$$(4.8) \qquad G_x'(z, x) = (z_3 - z_2 z_4 - z_3 z_4 x)e^{-z_4 x}.$$

We can observe that the mapping $G : \mathcal{Z} \longmapsto \mathcal{H}_\gamma$ is smooth and injective, its Fréchet derivative G_z' is injective and G_x' is smooth.

In the degenerate case $z_4 = 0$, we have

$$G(z, x) = z_1 + z_2 + z_3 x$$

and so z_2 is redundant. Thus we may set $z_2 = 0$ and obtain the forward curve family

$$G(z, x) = z_1 + z_3 x, \qquad (z_1, z_3) \in \mathbb{R}^2.$$

Let us start to analyze the consistency of the Nelson-Siegel family with the *Hull-White extended Vasiček model* (HW).

The Musiela parameterization of the Heath-Jarrow-Morton forward rate formulation of HW is given by

$$dr(t, x) = \alpha(t, x)\, dt + \sigma e^{-ax}\, dW(t).$$

Thus the volatility function is given by $\sigma(x) = \sigma e^{-ax}$, and the consistent drift and volatility conditions become

$$(4.9) \qquad G'_x(z, x) + \frac{\sigma^2}{a}\left[e^{-ax} - e^{-2ax}\right] \in \mathrm{Im}\left[G'_z(z, x)\right],$$

$$(4.10) \qquad \sigma e^{-ax} \in \mathrm{Im}\left[G'_z(z, x)\right].$$

We can immediately observe that (4.10) holds if and only if, for all $x \geq 0$, we have

$$(4.11) \qquad \sigma e^{-ax} = A + Be^{-z_4 x} + Cxe^{-z_4 x} - D(z_2 + z_3 x)xe^{-z_4 x},$$

and so if and only if $z_4 = a$. However, condition (4.10) must hold for all choices of z (see Theorem 4.4.3) and so HW is *inconsistent* with the manifold $\mathcal{G} = \mathrm{Im}[G]$ generated by NS.

Nevertheless we can modify NS in order to find a manifold which is consistent with HW. Because of (4.11), we can impose $z_4 = a$. The consistent drift condition is now

$$(z_3 - az_2 - az_3 x)e^{-ax} + \frac{\sigma^2}{a}\left[e^{-ax} - e^{-2ax}\right] = A + Be^{-ax} + Cxe^{-ax},$$

for all x, so we must expand the manifold by an exponential of the form e^{-2ax}. Thus we find the forward curve family defined by

$$G(x, z) = z_1 + z_2 e^{-ax} + z_3 x e^{-ax} + z_4 e^{-2ax}, \qquad z \in \mathbb{R}^4,$$

which is *consistent* with HW.

In the same way, we can easily test the consistency between NS and the *Ho-Lee model* (HL).

The Musiela parameterization of the Heath-Jarrow-Morton forward rate formulation of HL is given by

$$dr(t, x) = \beta(t, x)\, dt + \sigma\, dW(t).$$

Thus we have that NS is *inconsistent* with HL, while the degenerate family $G(z, x) = z_1 + z_3 x$ is *consistent* with HL.

4.6. – New forward curve families

In this section we present two new forward curve families of exponential-polynomial type. More details can be found in [2].

DEFINITION 4.6.1 (Exponential-polynomial family). *The forward curve manifold $EP(K, n)$, where K is a fixed positive integer and $n = (n_1, \ldots, n_K)$ is a vector of non-negative integers, is defined as the set of all curves of the form*

$$G(x) = \sum_{i=1}^{K} p_i(x) e^{-\alpha_i x},$$

where α_i is a non-negative real number for all i, and p_i is any polynomial with $\deg(p_i) \leq n_i$ for all i.

We have the following proposition:

PROPOSITION 4.6.2. *No non-trivial forward rate model with a deterministic volatility is consistent with the forward curve family $EP(K, n)$.*

In order to obtain a consistent forward rate model, we must fix some or all of the exponents $\alpha_1, \ldots, \alpha_K$. Thus we consider the following family:

DEFINITION 4.6.3 (Restricted family). *Consider a fixed choice of (K, n) and a vector $\beta = (\beta_1, \ldots, \beta_K)$ in $\mathbb{R}_+^K$. The restricted forward curve manifold $REP(K, n, \beta)$ is defined as the set of all curves of the form*

$$G(x) = \sum_{i=1}^{K} p_i(x) e^{-\beta_i x},$$

where p_i is any polynomial with $\deg(p_i) \leq n_i$ for all i.

We can observe that the consistent families for the interest rate models illustrated in the previous section are in this class: for Hull-White it is $REP(3, (0, 1, 0), (0, a, 2a))$ and for Ho-Lee it is $REP(1, 1, 0)$.

We have the following results:

PROPOSITION 4.6.4. *Consider a fixed family $REP(K, n, \beta)$ with no purely polynomial part (i.e. $\beta_i \neq 0$ for all i). Then a volatility $\sigma(x)$ is consistent with this family if and only if the following conditions are satisfied:*

(i) *The volatility σ must have the form*

$$\sigma(x) = \sum_{i=1}^{L} \widehat{p}_i(x) e^{-\beta_i x},$$

with $L \leq K$ and $\deg(p_i) \leq n_i$ for all i.

(ii) *For all i, j in $\{1, \ldots, L\}$ there exists an index k in $\{1, \ldots, K\}$ such that*

$$\beta_i + \beta_j = \beta_k,$$
$$\deg(\widehat{p}_i) + \deg(\widehat{p}_j) \leq n_k.$$

COROLLARY 4.6.5. *Under the previous assumptions there exists a consistent (with $REP(K, n, \beta)$) volatility σ if and only if there exist two indices $i, k \leq K$ such that $2\beta_i = \beta_k$.*
If this condition is satisfied, any σ of the form $\sigma(x) = \widehat{p}_i e^{-\beta_i x}$, with $\deg(p_i) \leq n_i$ and $2\deg(p_i) \leq n_k$, will be consistent.

4.7. – The Filipoviĉ approach

In this section we illustrate an alternative approach, due to Filipoviĉ, concerning the consistency problem. The reader can find more details and results in [8] and [9].

Let us consider a given submanifold $\mathcal{G} = \text{Im}[G]$ of forward rate curves, with $G : \mathbb{R}^m \longmapsto \mathcal{H}$. We want to investigate when G is supported by an arbitrage free interest rate model. More formally, we want to know if there exists some m-dimensional stochastic differential equation of the form

$$dZ_t = a(Z_t)\,dt + b(Z_t)\,dW_t$$

such that the process r_t defined by

$$r_t(x) = G(x; Z_t)$$

is an arbitrage free forward rate model, in the sense that r_t satisfies the Heath-Jarrow-Morton-Musiela (HJMM) equation given by

$$dr_t(x) = \left\{ \frac{\partial}{\partial x} r_t(x) + \sigma(t, x) \int_0^x \sigma(t, s)^\star \, ds \right\} dt + \sigma(t, x)\,dW(t)$$

(where the symbol $\star$ denotes transpose).

Using Itô formula, we find

$$dr_t = \left\{ G_z(Z_t)a(Z_t) + \frac{1}{2}\text{tr}\left[b^\star(Z_t)G_{zz}(Z_t)b(Z_t) \right] \right\} dt$$

$$+ G_z(Z_t)b(Z_t)dW_t.$$

Comparing (4.12) with HJMM equation, we can rewrite our problem in the following way:

Consider a given submanifold $\mathcal{G} = \text{Im}[G]$ of forward rate curves, with $G : \mathbb{R}^m \longmapsto \mathcal{H}$, we ask when there exist a vector field $a(z)$ and a matrix field $b(z)$ such that

$$G_z(x, z)a(z) + \frac{1}{2}\text{tr}\left[b^\star(z)G_{zz}(x, z)b(z) \right]$$

$$= G_x(x, z) + G_z(x, z)b(z) \int_0^x (G_z(s, z)b(z))^\star \, ds.$$

The main results by Filipoviĉ, concerning this question, are the following:

1. No non-deterministic forward rate model is consistent with Nelson-Siegel family.

2. No non-deterministic forward rate model is consistent with an exponential-polynomial family with varying exponents.

CHAPTER V

Finite Dimensional Realizations of Forward Rate Models

In this chapter we consider the problem when a given forward rate model with a time-invariant volatility, has a finite dimensional realization. For more precise results we refer to [4], where it is possible to find also a discussion on the time-varying case.

5.1. – Some differential geometry

In this section we give the background on infinite dimensional differential geometry necessary and sufficient for the sequel.

Let X, Y be real Banach spaces.

DEFINITION 5.1.1. *Let U, V be open subsets respectively of X and Y. A smooth map $\varphi : U \longmapsto V$ is a* **diffeomorphism** *if it has a smooth inverse.*

(The prefix "smooth" above is interpreted as C^∞.)

DEFINITION 5.1.2. *Let $\mathcal{Z}$ be an open connected subset of a Banach space. A* **submanifold** *G in X is a mapping $G : \mathcal{Z} \longmapsto X$ such that G is injective, its Fréchet derivative $G'(x)$ is injective for all x and $\mathcal{G} = \text{Im}[G]$ is homeomorphic to $\mathcal{Z}$.*

In the sequel we will often call "submanifold" the image, $\mathcal{G}$, of G.

DEFINITION 5.1.3. *Let us consider a submanifold G and a point x in*
$\mathcal{G} = \text{Im}[G]$. *Moreover let* $v : (-1, 1) \longmapsto \mathcal{G}$ *be any smooth curve with*
$v(0) = x$. *Then we say that the vector*

$$h = \dot{v}(0) = \frac{dv}{dt}(0)$$

is a **tangent vector** *to* $\mathcal{G}$ *at x. The space of all tangent vectors to* $\mathcal{G}$ *at x is*
denoted by $T_{\mathcal{G}}(x)$.

DEFINITION 5.1.4. *Let U be an open subset of X. A* **vector field** *on U*
is a smooth mapping

$$f : U \longmapsto X.$$

If $\varphi : U \longmapsto V$ is a diffeomorphism between two open subsets of
X and f is a vector field on U, we can define a new vector field $\varphi_* f$
on V setting

(5.1) $$\varphi_* f = \left(\varphi' \circ \varphi^{-1} \right) \left(f \circ \varphi^{-1} \right).$$

Let us consider a vector field f on X and a point a in X. If we
want to find a finite dimensional submanifold $\mathcal{G} \subseteq X$ which contains
a and which is tangential to f, that is a submanifold satisfying the
following conditions:

(i) $a \in \mathcal{G}$,

(ii) $f(x) \in T_{\mathcal{G}}(x)$ for all x in a neighborhood of a in $\mathcal{G}$,

then we have to solve the ordinary differential equation

(5.2)
$$\begin{cases} \dot{x}(t) = f(x(t)) \\ \\ x(0) = a. \end{cases}$$

We denote the solution of equation above by $x_t(a)$ or by $e^{ft}a$. The
manifold $t \longmapsto x_t(a) = e^{ft}a$ is called the **integral curve** of f through a.

Now, let us take as given two vectors fields f, g on X and a point a
in X. We want to investigate if there is a finite dimensional submanifold
$\mathcal{G}$ with a in $\mathcal{G}$ and which is tangential to both f and g in the sense
that Span $\{f(x), g(x)\}$ is contained in $T_{\mathcal{G}}(x)$ for all x in a neighborhood

of a in $\mathcal{G}$. In order to answer to this question, we give the following definition:

DEFINITION 5.1.5. *Given two vector fields f and g defined on the open subset U, their* **Lie bracket** *$[f, g]$ is the vector field on U defined by*

$$[f, g](x) = f'(x)g(x) - g'(x)f(x),$$

where $f'(x)g(x)$ (resp. $g'(x)f(x)$) denotes the Fréchet derivative of f (resp. g) at x operating on $g(x)$ (resp. $f(x)$).

It easy to see that the Lie bracket operator is bilinear over $\mathbb{R}$ and that, if α is a scalar field, we have

$$[f, \alpha \cdot g] = \alpha \cdot [f, g] - (\alpha' f) \cdot g.$$

Moreover, if $\varphi : U \longmapsto V$ is a diffeomorphism, then we have the following relation:

(5.3) $$\varphi_*[f, g] = [\varphi_* f, \varphi_* g].$$

Now we are ready to answer our question.

PROPOSITION 5.1.6. *A necessary and sufficient condition for the existence of a two-dimensional tangential manifold to both f and g is the following:*

$$[f, g](x) \in \text{Span}\{f(x), g(x)\}, \quad \text{for all } x.$$

This proposition is a consequence of Frobenius Theorem illustrated below.

DEFINITION 5.1.7. *Let V be an open subset of X. A* **distribution** *on V is a mapping F on V such that, for each x in V, $F(x)$ is a subspace of X. The* **dimension** *of a distribution F is defined pointwise as $\dim F(x)$.*

If a collection $\{f_1, \ldots, f_n\}$ of smooth vector fields on V is such that, for every x in V, we have

$$F(x) = \text{Span}\{f_1(x), \ldots, f_n(x)\},$$

we say that the distribution F is **generated** *by $f_1, \ldots, f_n$ on V.*

DEFINITION 5.1.8. *Let F be a distribution on V. A vector field f on U is said to* **lie in** *F (on U), if U is a subset of V and $f(x)$ is an element of $F(x)$, for every x in U.*

A distribution F is called **involutive** *if, for every vector fields f and g lying in F, their Lie bracket $[f, g]$ also lies in F.*

Let F be a distribution on V. A diffeomorphism $\varphi : V \longmapsto W$ induces a new distribution $\varphi_* F$ on W defined by

$$(\varphi_* F)(\varphi(x)) = \varphi'(x) F(x)$$

or, equivalently,

$$(\varphi_* F)(y) = \varphi'\left(\varphi^{-1}(y)\right) F\left(\varphi^{-1}(y)\right).$$

REMARK 5.1.9. From relations (5.1) and (5.3) we obtain that, if F is generated by $f_1, \ldots, f_n$, then $\varphi_* F$ is generated by $\varphi_* f_1, \ldots, \varphi_* f_n$ and F is involutive if and only if $\varphi_* F$ is involutive.

THEOREM 5.1.10 (Frobenius). *Let F be a smooth distribution on an open subset V of X, that is a distribution F on V such that, for each x in V, there exist smooth vector fields $f_1, \ldots, f_n$ generating F on a neighborhood of x. Further, let us fix an arbitrary point x in V. Then there exists a diffeomorphism $\varphi : U \longmapsto X$ on some neighborhood $U \subseteq V$ of x such that $\varphi_* F$ is constant on $\varphi(U)$ if and only if F is involutive.*

PROOF. The "only if part" is a consequence of the remark above.

Let us assume that F is involutive. In order to prove the existence of φ, we proceed by induction on n. Thus let us suppose $n = 1$. Without loss of generality we may assume that 0 is an element of V and take $x = 0$. Moreover we may suppose that F is generated by a smooth vector field f on V. Let us define the vector e by $e = f(0)$ and, since $\mathbb{R} e$ is a finite dimensional subspace in a Banach space, we can write

$$X = \mathbb{R} e \oplus Y.$$

Now, for each fixed y in Y, let us consider the equation

$$\begin{cases} \dot{z}(t) & = f(z(t)) \\ \\ z(0) & = y. \end{cases}$$

Because of the existence theorem for ordinary differential equation, we find a unique smooth local solution $z(t, y)$ of equation above. Setting

$$\psi(te + y) = z(t, y),$$

we have $\psi(y) = y$ and

$$\psi'(te + y)e = \dot{z}(t, y) = f(z(t, y)) = f(\psi(te + y)).$$

We also have the equality

$$\psi(te + y) = y + \int_0^t f(\psi(se + y)) \, ds.$$

From equality above, we obtain, for small t and y near 0,

$$\psi(te + y) = y + \int_0^t f(\psi(0)) + O(se + y) \, ds$$
$$= y + tf(0) + o(te + y)$$
$$= y + te + o(te + y).$$

Thus we have $\psi'(0)(te + y) = tf(0) + y = te + y$, that is $\psi'(0)$ is the identity mapping. Then the inverse function theorem gives us a local inverse $\varphi = \psi^{-1}$ defined on a neighborhood U of 0 in X.

Now let $x = \psi(te + y)$ (with x in U), that is $\varphi(x) = te + y$. It follows that

$$(\varphi_* f)(te + y) = \varphi'(x) f(x).$$

On the other hand, we have

$$\psi'(te + y)e = f(\psi(te + y)) = f(x),$$

and so

$$(\varphi_* f)(te + y) = \varphi'(\psi(te + y))\psi'(te + y)e = (\varphi \circ \psi)'(te + y)e = e,$$

that is $\varphi_* f$ coincides with the constant vector field e on $\varphi(U)$.

For the induction step, let us consider n with $n > 1$ and let us suppose that the theorem holds for every m-dimensional distribution with $m < n$. Without loss of generality, we may again assume that 0 is an element of V and take $x = 0$. Further, let us assume that $f_1, \ldots, f_n$ generate F on V. Let us set $e_i = f_i(0)$ and, since $\mathbb{R}e_1 + \cdots \mathbb{R}e_n$ is a finite dimensional subspace in a Banach space, we can find a close subspace Z of X such that

$$X = \mathbb{R}e_1 + \cdots \mathbb{R}e_n + Z$$

$$(\mathbb{R}e_1 + \cdots \mathbb{R}e_n) \cap Z = 0$$

Now let us define the subspace Y by

$$Y = \mathbb{R} e_2 + \cdots \mathbb{R} e_n + Z$$

Since the theorem holds for $n = 1$, we may (modulo a diffeomorphism) assume that $f_1 = e_1$. Moreover, by applying Gauss elimination (that is dividing by scalar functions and subtracting), we may also assume that

$$f_j = e_j + g_j, \quad \text{for } j = 2, \ldots, n,$$

with g_j in $C^\infty(V, Z)$.

Since F is supposed involutive, there exist scalar fields $\alpha_{j,k}$ in $C^\infty(V, \mathbb{R})$ such that

$$[f_1, f_j] = \sum_{k=1}^{n} \alpha_{j,k} f_k \quad \text{for } j = 2, \ldots, n,$$

and so we have

$$[f_1, f_j] = f_1' f_j - f_j' f_1 = 0 - g_j' e_1$$
$$= \alpha_{j,1} e_1 + \cdots + \alpha_{j,n} e_n + \alpha_{j,2} g_2 + \cdots + \alpha_{j,n} g_n.$$

Now we observe that g_j' is a mapping defined on V and with values in $L(X, Z)$ (where $L(X, Z)$ denotes the space of bounded linear maps on X with values in Z) and so $g_j' e_1$ has values in Z. It follows that, for $j = 2, \ldots, n$, we have $\alpha_{j,1} = \cdots = \alpha_{j,n} = 0$. So $g_j' e_1 = 0$ and we have $g_j(t_1 e_1 + y) = g_j(y)$ for every t_1 in $\mathbb{R}$ and y in Y. Therefore the restrictions of $f_2, \ldots, f_n$ to Y generates an $(n - 1)$-dimensional distribution F_Y which is smooth and involutive. From our induction hypothesis, we deduce the existence of a diffeomorphism φ_Y, defined on a neighborhood of 0 in Y, such that $(\varphi_Y)_* F_Y$ is constant. Let φ be the extension of φ_Y define by

$$\varphi(t e_1 + y) = t e_1 + \varphi_Y(y).$$

Then φ is a diffeomorphism in a neighborhood U of 0 in X such that $\varphi_* F$ is constant on $\varphi(U)$.   □

We use Frobenius Theorem in order to prove the existence of tangential manifolds.

DEFINITION 5.1.11. *Let F be a smooth distribution, and let a be a fixed point in X. A submanifold $\mathcal{G} \subseteq X$ with a in $\mathcal{G}$ is called a* **tangential manifold** *through a for F, if $F(x)$ is contained in $T_\mathcal{G}(x)$ for each x in a neighborhood of a in $\mathcal{G}$.*

We have the following corollary:

COROLLARY 5.1.12. *Let F be a smooth distribution on X and a a point in X. Then the following statements are equivalent:*

(i) *For every x in a neighborhood of a, there exists a finite dimensional tangential manifold $\mathcal{G}$ for F passing through x.*

(ii) *The distribution F is involutive.*

For the proof see [4].

Now we give a new definition:

DEFINITION 5.1.13. *Let F be a smooth distribution. The* **Lie algebra** *generated by F, denoted by $\{F\}_{LA}$, is defined as the minimal (under inclusion) involutive distribution containing F. If F is generated by the vectors fields $f_1, \ldots, f_n$, we will denote $\{F\}_{LA}$ by $\{f_1, \ldots, f_n\}_{LA}$.*

If, for example, the distribution F is generated by the vector fields $f_1, \ldots, f_n$, then, in order to construct the Lie algebra $\{f_1, \ldots, f_n\}_{LA}$, we have to form all possible brackets and brackets of brackets, etc. of vector fields $f_1, \ldots, f_n$, and adjoin these to the original distribution until the dimension of the distribution doesn't increase any longer.

From Frobenius Theorem, we have the following corollary:

COROLLARY 5.1.14. *Let F be a smooth distribution on X and a a point in X. Then the following statements are equivalent:*

(i) *For every x in a neighborhood of a, there exists a finite dimensional tangential manifold $\mathcal{G}$ for F passing through x.*

(ii) *We have $\dim\{F\}_{LA}(x) < \infty$ for each x in a neighborhood of a.*

Finally, we have the following results:

PROPOSITION 5.1.15. *Let us consider a smooth distribution F. If F generates a finite dimensional Lie algebra $\{F\}_{LA}$, and if $\{F\}_{LA}$ is generated by $\{f_1, \ldots, f_n\}$, then the tangential manifold $\mathcal{G}$ for F passing through a can be parameterized as*

$$G(z_1, \ldots, z_n) = e^{f_n z_n} \ldots e^{f_1 z_1} a.$$

PROPOSITION 5.1.16. *Take the vector fields $f_1, \ldots, f_n$ as given. The Lie algebra $\{f_1, \ldots, f_n\}_{LA}$ remains unchanged under the following operations.*

- *The vector field f_i may be replaced by αf_i, where α is any smooth nonzero scalar field.*

- *The vector field f_i may be replaced by*

$$f_i + \sum_{j \neq i} \alpha_j f_j,$$

where each α_j is any smooth scalar field.

5.2. – The space

As in the previous chapter, we denote the forward rate curve at time t by r_t and we use the compact notation $\sigma(r)$ for $\sigma(r, x)$. Moreover we assume that the Q-dynamics of the forward rate process is

$$dr_t = \left\{ \frac{\partial}{\partial x} r_t + \sigma(r_t) H \sigma(r_t)^\star \right\} dt + \sigma(r_t) dW_t,$$

$$r_0(x) = r_0^*(x)$$

where W is a d-dimensional Wiener process and

$$H\sigma(r, x)^\star = \int_0^x \sigma(r, s)^\star \, ds.$$

(The symbol $\star$ denotes transpose.)

Now we observe that in most "naturally" chosen spaces, the operator $\frac{\partial}{\partial x}$ is unbounded, which means that we have no existence results concerning strong solutions. Thus we need a very regular space to work in.

DEFINITION 5.2.1. *Let us fix $\alpha > 0$ and $\beta > 1$. The space $\mathcal{H}_{\alpha,\beta}$ is defined by*

$$\mathcal{H}_{\alpha,\beta} = \{ f \in C^\infty[0, \infty) \, ; \, \|f\|_{\alpha,\beta} < \infty \},$$

where

$$\|f\|_{\alpha,\beta}^2 = \sum_{n=0}^{\infty} \beta^{-n} \int_0^\infty \left[\frac{d^n f}{dx^n}(x) \right]^2 e^{-\alpha x} \, dx.$$

In the sequel, we suppress the subindices and we write $\mathcal{H}$ instead of $\mathcal{H}_{\alpha,\beta}$.

The space $\mathcal{H}$ has the following properties:

PROPOSITION 5.2.2.

(i) *With the inner product defined by*

$$(f, g) = \sum_{n=0}^{\infty} \beta^{-n} \int_0^{\infty} \left[\frac{d^n f}{dx^n}(x) \right]^2 \left[\frac{d^n g}{dx^n}(x) \right]^2 e^{-\alpha x} \, dx,$$

the space $\mathcal{H}$ is a Hilbert space.

(ii) *The linear operator $F = \frac{\partial}{\partial x}$ is bounded on $\mathcal{H}$.*

(iii) *The elements in $\mathcal{H}$ are real analytic functions and can be extended to analytic functions on the entire real line $\mathbb{R}$.*

We then take as given a volatility σ of the form

$$\sigma : \mathcal{H} \times \mathbb{R}_+ \longmapsto \mathbb{R}^d,$$

that is each component of $\sigma(r, x) = [\sigma_1(r, x), \ldots, \sigma_d(r, x)]$ is a functional of the infinite dimensional r-variable and a function of the real variable x.

In the sequel we also need the following assumptions:

ASSUMPTIONS. *We see each component σ_i of σ as a mapping from $\mathcal{H}$ to a space of functions and we assume that σ has the following properties:*

1. *The mappings $\sigma_1, \ldots, \sigma_d$ are smooth vector fields on $\mathcal{H}$.*

2. *The mapping $r \longmapsto \sigma(r)H\sigma(r)^*$ is a smooth vector field on $\mathcal{H}$.*

We recall (see equation (4.3)) that the Stratonovich form of the forward rate model (5.4) is given by

(5.5) $\qquad dr_t = \alpha(r_t) \, dt + \sigma(r_t) \circ dW_t, \qquad r_0 = r^*,$

where

(5.6) $\qquad \alpha(r) = \frac{\partial}{\partial x} r + \sigma(r)H\sigma(r)^* - \frac{1}{2}\sigma'(r)\sigma(r).$

Using the Stratonovich differential, we can treat the stochastic differential equation (5.5) as an equation of the form

$$\frac{dr_t}{dt} = \alpha(r_t) + \sigma(r_t) \cdot v_t,$$

where v_t is interpreted as "white noise".

5.3. – Finite dimensional realizations

In this section we investigate under what conditions the process r can be realized by a finite dimensional stochastic differential equation.

DEFINITION 5.3.1. *We say that the stochastic differential equation (5.5) has a* **local m-dimensional realization** *if there exist a point z_0 in $\mathbb{R}^m$, smooth vector fields $a, b_1, \ldots, b_d$ on some open connected subset $\mathcal{Z}$ of $\mathbb{R}^m$ and a smooth map $G : \mathcal{Z} \longmapsto \mathcal{H}$, such that r has the local representation*

$$(5.7) \qquad\qquad\qquad r_t = G(Z_t)$$

where Z is the solution of the m-dimensional Stratonovich stochastic differential equation

$$(5.8) \qquad dZ_t = a(Z_t)\, dt + b(Z_t) \circ dW_t \qquad Z_0 = z_0.$$

The prefix "local" above means that the representation is assumed to hold for all t with $0 \le t < \tau(r^*)$, Q-a.s. where, for each r^* in $\mathcal{H}$, $\tau(r^*)$ is a strictly positive stopping time.

We can observe that the realization concept is closely connected to the concept of locally invariant (or r-invariant) submanifold (see Chapter 4). More precisely, we have the following theorem:

THEOREM 5.3.2. *There exists a local finite dimensional realization of (5.5) if and only if there exists a finite dimensional locally invariant manifold $\mathcal{G}$ with r^* in $\mathcal{G}$.*

For the proof see [4].

The problem of finding a realization is thus reduced to the problem of finding a finite dimensional invariant submanifold. Using Theorem 4.4.3., we can link this problem to Frobenius theory illustrated in Section 5.1.

THEOREM 5.3.3. *A manifold $\mathcal{G}$ is locally invariant under r if and only if the conditions*

$$\alpha(r) \in T_{\mathcal{G}}(r)$$
$$\sigma(r) \in T_{\mathcal{G}}(r)$$

hold at all points $r = G(z)$ of $\mathcal{G}$. Thus $\mathcal{G}$ is locally invariant if and only if it is a tangential manifold for the distribution generated by the vector fields $\alpha, \sigma_1, \ldots, \sigma_d$.

Summing up, we have the following main result:

THEOREM 5.3.4. *The stochastic differential equation (5.5) has a local finite dimensional realization if and only if there exists a finite dimensional tangential manifold for the distribution generated by the vector fields* $\alpha, \sigma_1, \ldots, \sigma_d$, *containing the initial point* r^*.

Finally, by results of Section 5.1, we have the following theorem on the existence of finite dimensional realizations:

THEOREM 5.3.5. *Let us take as given a forward rate curve* $\hat{r}$ *in* $\mathcal{H}$. *The following statements are equivalent:*

(i) *For each choice of initial point* r^* *in a neighborhood of* $\hat{r}$ *in* $\mathcal{H}$, *there exists a local finite dimensional realization of (5.5).*

(ii) *The Lie algebra* $\mathcal{L} = \{\alpha, \sigma\}_{LA} = \{\alpha, \sigma_1, \ldots, \sigma_d\}_{LA}$ *is finite dimensional in a neighborhood of* $\hat{r}$.

The dimension of the realization is equal to the dimension of $\mathcal{L}$.

REMARK 5.3.6. Realizations are not unique: any diffeomorphism of the parameter space $\mathcal{Z} \subseteq \mathbb{R}^m$ will give an equivalent realization. Moreover, we can find realizations where the state variables have an economic interpretation. More precisely, it is proved (see [4]) that for a minimal (with respect to the dimension m) realization we can always choose the states as a set of *benchmark forward rates*.

5.4. – Deterministic volatility

In this section we consider the case when the volatility is a smooth deterministic function $\sigma(x)$ with values in $\mathbb{R}^d$. In this case the Itô and Stratonovich formulations of the forward rate model are the same and we can write

$$dr = \{Fr + D\} + \sigma\, dW,$$

where $F = \frac{\partial}{\partial x}$ and D is the function defined by

$$D(x) = \sigma(x) \int_0^x \sigma(s)\, ds.$$

We set $\alpha = Fr + D$.

Since D and σ are constant vector fields on $\mathcal{H}$, we have that the Lie bracket $[\alpha, \sigma_i]$ is given by

$$[Fr + D, \sigma_i] = F\sigma_i.$$

For the same reason, we have

$$[Fr + D, [Fr + D, \sigma_i]] = F^2\sigma_i$$

and, continuing in the same way, we find

$$\mathcal{L} = \{\alpha, \sigma_1, \ldots, \sigma_d\}_{LA} = \{Fr + D, \sigma_1, \ldots, \sigma_d\}_{LA}$$
$$= \text{Span}\,[Fr + D, F^k\sigma_i \,;\, i = 1, \ldots, d \quad k \geq 0].$$

By Theorem 5.3.5, we find again the following result:

THEOREM 5.4.1. *Let us assume that the volatility is a smooth deterministic function with values in $\mathbb{R}^d$. Then the model has a finite dimensional realization if and only if the space*

$$\text{Span}\,[\,F^k\sigma_i \,;\, i = 1, \ldots, d \quad k \geq 0\,]$$

is finite dimensional. (See also Proposition 3.5.2.)

Therefore, we have a finite dimensional realization if and only if σ solves an ordinary differential equation with constant coefficients, that is if and only if σ can be written in the form

$$\sigma(x) = Ce^{Ax}B,$$

where A, B are matrices and C is a row-vector (see also Proposition 3.4.4).

5.5. – A concrete example

In this section we illustrate in a special case how to construct a locally invariant submanifold $\mathcal{G}$ with r^* in $\mathcal{G}$.

Let us take as given a forward curve r^* and let us assume, for simplicity, $d = 1$. We also suppose that the volatility σ is a deterministic function and (using the notation of the previous section)

$$\mathcal{L} = \{Fr + D, \sigma\}_{LA} = \text{Span}\,[Fr + D, \sigma, F\sigma, \ldots, F^n\sigma\,].$$

Then it can be proved (see Proposition 5.1.15) that the tangential manifold $\mathcal{G}$ for the distribution generated by $Fr + D$ and σ, passing through r^*, is parameterized as

$$(5.9) \qquad G(t, t_0, t_1, \ldots, t_n) = e^{F^n t_n} \ldots e^{F t_1} e^{\sigma t_0} e^{(Fr+D)t} r^*,$$

where $t \longmapsto e^{(Fr+D)t} r^*$ solves in $\mathcal{H}$ the ordinary differential equation

$$\begin{cases} \frac{dr}{dt} &= Fr + D \\ r(0) &= r^* \end{cases}$$

and, for each $k = 0, \ldots, n$, the curve $t \longmapsto e^{F^k t} r_0$ solves in $\mathcal{H}$ the ordinary differential equation

$$\begin{cases} \frac{dr}{dt} &= F^k \sigma = \frac{d^k \sigma}{dx^k} \\ r(0) &= r_0. \end{cases}$$

Thus we have

$$(5.10) \qquad \left(e^{(Fr+D)t} r^* \right)(x) = r^*(x + t) + \int_0^t D(x + t - s) \, ds$$

and

$$(5.11) \qquad \left(e^{F^k t} r_0 \right)(x) = r_0(x) + t \frac{d^k \sigma}{dx^k}(x) \quad k = 0, \ldots, n.$$

Putting (5.10) and (5.11) in (5.9), we finally find

$$G(t, t_0, t_1, \ldots, t_n)(x) = r^*(x + t) + \int_0^t D(x + t - s) \, ds + \sum_{k=0}^n t_k \frac{d^k \sigma}{dx^k}(x).$$

Now let us also take as given a finite dimensional manifold $\mathcal{G}_0$ which is not locally invariant. It can be proved (see [4]) that it is possible to extend $\mathcal{G}_0$ to a finite dimensional locally invariant manifold. More precisely, we have the following result:

PROPOSITION 5.5.1. *Let us consider a fixed interest rate model admitting a finite dimensional locally invariant manifold and let us take as given a manifold $\mathcal{G}_0$, parameterized by $G_0(z_1, \ldots, z_k)$, which is not locally invariant. Moreover let us suppose that the Lie algebra $\mathcal{L} = \{\alpha, \sigma\}_{LA}$ is generated by* *the vector fields $\alpha, f_1, \ldots, f_n$ and that $\mathcal{L}$ is transversal to $\mathcal{G}_0$. Then $\mathcal{G}_0$ can be minimally extended to a locally invariant submanifold $\mathcal{G}$, where $\mathcal{G}$ is parameterized by the mapping*

$$G(z_1, \ldots, z_k, t_0, t_1, \ldots, t_n) = e^{f_n t_n} \ldots e^{f_1 t_1} e^{\alpha t_0} G_0(z_1, \ldots, z_k).$$

REMARK 5.5.2. The term "transversal" above means that no vectors in the Lie algebra $\mathcal{L}$ is contained in the tangent space of $\mathcal{G}_0$ at any point of $\mathcal{G}_0$.

EXAMPLE 5.5.3 Let us consider the Hull-White extended Vasiček model (HW) that is $\sigma(x) = \sigma e^{-ax}$. Then we have

$$D(x) = \frac{\sigma^2}{a} e^{-ax} \left[1 - e^{-ax}\right]$$

and

$$\text{Span}\left[F^k \sigma ; k \geq 0\right] = \text{Span}\{\sigma e^{-ax}\} = \text{Span}\{e^{-ax}\}.$$

Therefore, for a given initial forward rate curve r^*, the locally invariant manifold generated by HW is parameterized by

$$G(t, t_0)(x) = r^*(x+t) + \frac{\sigma^2}{a^2} e^{-ax}\left[1 - e^{-at}\right] - \frac{\sigma^2}{2a^2} e^{-2ax}\left[1 - e^{-2at}\right] + t_0 e^{ax}.$$

If we take as given the Nelson-Siegel manifold $\mathcal{G}_0$, parameterized as

$$G_0(z_1, \ldots, z_4)(x) = z_1 + z_2 e^{-z_4 x} + z_3 x e^{-z_4 x},$$

then the minimal locally invariant extension $\mathcal{G}$ is parameterized by

$$G(t, t_0, z_1, \ldots, z_4)(x) = e^{\sigma t_0} e^{(Fr+D)t} G_0(z_1, \ldots, z_4)(x)$$

$$= G_0(z_1, \ldots, z_4)(x+t) + \int_0^t D(x+t-s)ds + t_0 e^{-ax}$$

$$= G_0(z_1, \ldots, z_4)(x+t) + \frac{\sigma^2}{a^2} e^{-ax}\left[1 - e^{-at}\right]$$

$$- \frac{\sigma^2}{2a^2} e^{-2ax}\left[1 - e^{-2at}\right] + t_0 e^{ax}.$$

Since $\mathcal{G}_0$ is invariant under shift in x, then a simpler parameterization of $\mathcal{G}$ is

$$G(t, t_0, z_1, \ldots, z_4)(x) = G_0(z_1, \ldots, z_4)(x) + \frac{\sigma^2}{a^2} e^{-ax} \left[1 - e^{-at} \right]$$

$$- \frac{\sigma^2}{2a^2} e^{-2ax} \left[1 - e^{-2at} \right] + t_0 e^{ax}$$

$$= z_1 + z_2 e^{-z_4 x} + z_3 x e^{z_4 x} + \frac{\sigma^2}{a^2} e^{-ax} \left[1 - e^{-at} \right]$$

$$- \frac{\sigma^2}{2a^2} e^{-2ax} \left[1 - e^{-2at} \right] + t_0 e^{ax}.$$

5.6. – Constant direction volatility

We now consider a natural extension of the deterministic volatility case (still in the case of one scalar driving Wiener process, i.e. $d = 1$), namely the case when the volatility is of the form

$$\sigma(r, x) = \varphi(r)\lambda(x),$$

where φ is a smooth functional on $\mathcal{H}$ and λ an element of $\mathcal{H}$.

We assume the following:

1. $\varphi(r) \neq 0$ for all r in $\mathcal{H}$.
2. $\phi''(r)[\lambda; \lambda] \neq 0$ for all r in $\mathcal{H}$, where $\phi(r) = \varphi^2(r)$ and $\phi''(r)[\lambda; \lambda]$ is the second order Fréchet derivative of ϕ operating on the vector pair $[\lambda; \lambda]$.

Under these assumptions we have the following result:

PROPOSITION 5.6.1. *The model with volatility $\sigma(r, x) = \varphi(r)\lambda(x)$ has a finite dimensional realization if and only if λ is a quasi-exponential function, that is*

$$\lambda(x) = ce^{Ax}b,$$

where A is a square matrix, b a column-vector and c a row-vector. The scalar field φ is allowed to be any smooth field.

PROOF. We have $\sigma(r, x) = \varphi(r)\lambda(x)$ and so we find

$$\alpha(r) = Fr + \varphi^2(r)D - \frac{1}{2}\varphi'(r)[\lambda]\varphi(r)\lambda,$$

where $\varphi'(r)[\lambda]$ denotes the Fréchet derivative $\varphi'(r)$ operating on λ, and where D is defined by

$$D(x) = \lambda(x) \int_0^x \lambda(s)\, ds.$$

It is easy to see (using Proposition 5.1.16) that the Lie algebra $\mathcal{L}$ generated by α and σ coincides with the Lie algebra generated by the vector fields

$$f_0(r) = Fr + \phi(r)D$$

$$f_1(r) = \lambda$$

(where $\phi(r) = \varphi^2(r)$).

Now we can observe that

$$[f_0, f_1](r) = F\lambda + \phi'(r)[\lambda]D$$

$$[[f_0, f_1], f_1](r) = \phi''(r)[\lambda; \lambda]D.$$

Using again Proposition 5.1.16 we see that the Lie algebra $\mathcal{L}$ is generated by the following vector fields:

$$f_0(r) = Fr$$
$$f_1(r) = \lambda$$
$$f_2(r) = F\lambda$$
$$f_3(r) = D.$$

After elementary calculations, we obtain

$$\mathcal{L} = \{\alpha, \sigma\}_{LA} = \text{Span}\{Fr, F^n\lambda, F^nD\,;\, n \geq 0\}.$$

Therefore $\mathcal{L}$ is finite dimensional only if the vector space generated by $\{F^n\lambda; n \geq 0\}$ is finite dimensional. This occurs if and only if λ is quasi-exponential. On the other hand, if λ is quasi-exponential, then also D is quasi-exponential and so also the space generated by $\{F^nD; n \geq 0\}$ is finite dimensional. The proof is so concluded. $\square$

5.7. – Short rate realizations

In this section we illustrate how the theory developed above can be used in order to determine when a given forward rate model is realized by a short rate model, that is when the short rate model induced by the forward rate model is a Markov process. In order to avoid complications, we consider only one scalar driving Wiener process that is $d = 1$.

Let us suppose that the model is really a short rate model and that the bond prices are given as $p(t, T) = F^T(t, R_t)$ where F^T solves the term structure equation. We then have

$$r(t, x) = G(t, R_t, x),$$

where

$$G(t, R, x) = -\frac{\partial}{\partial x} \log F^{t+x}(t, R).$$

Hence we see that the forward rate process has a 2-dimensional realization with state process $Z = [t, R_t]$. Such a realization is termed **short rate realization**.

Using the results above, we immediately have the following necessary condition:

PROPOSITION 5.7.1. *The forward rate model generated by $\sigma(r, x)$ has a short rate realization (i.e. the short rate is a Markov process), only if*

$$\dim \{\alpha, \sigma\}_{LA} \leq 2.$$

Let us assume that $\dim\{\alpha, \sigma\}_{LA} = 2$. This condition implies that there exists a 2-dimensional realization but we don't know whether the realization can be chosen in such a way that the short rate R and the running time t are the state variables. This problem is completely solved by the following result:

THEOREM 5.7.2. *Assuming $\dim\{\alpha, \sigma\}_{LA} = 2$ there exists a short rate realization if and only if the vector fields $[\alpha, \sigma]$ and σ are parallel, that is if and only if there exists a scalar field $\varphi(r)$ such that the relation*

$$[\alpha, \sigma](r) = \varphi(r)\sigma(r)$$

holds locally for all r.

REMARK 5.7.3. The case dim$\{\alpha, \sigma\}_{LA} = 2$ is the most natural case. However, it is an open problem whether there exists a nontrivial model with dim$\{\alpha, \sigma\}_{LA} = 1$.

We conclude with the following result, which was first proved by Jeffrey [11]. See [4] for a proof based on the previous theorem.

THEOREM 5.7.4. *Let us assume that the forward rate volatility is of the form*

$$\sigma(r, x) = \sigma(R, x),$$

where $R = r(0)$ is the short rate. Then the model has a short rate realization if and only if it is **affine**, *that is if and only if $\sigma(R, x)$ has one of the following forms:*

(i) *There exists a constant σ such that*

$$\sigma(R, x) = \sigma$$

(Ho-Lee model).

(ii) *There exist constants a and σ such that*

$$\sigma(R, x) = \sigma e^{-ax}$$

(Hull-White extended Vasiček model).

(iii) *There exist constants a and b, and a function $\varphi(x)$, where φ satisfies a certain Riccati equation, such that*

$$\sigma(R, x) = \varphi(x)\sqrt{aR + b}$$

(Hull-White extended Cox-Ingersoll-Ross model).

In order to understand better the result above, let us consider the following program:

1. Choose an arbitrary short rate model, say of the form

$$dR_t = a(t, R_t)\, dt + b(R_t)\, dW_t$$

with a fixed initial point R_0.

2. Solve the associated partial differential equation in order to compute bond prices. This will also produce:
   * An initial forward rate curve $\hat{r}(x)$
   * Forward rate volatilities of the form $\sigma(R, x)$.

3. Forget about the underlying short rate model, and take the forward rate volatility structure $\sigma(R, x)$ as given.

4. Initiate this forward rate equation with an arbitrary initial forward rate curve $r^*(x)$.

Now we ask whether the so constructed forward rate model will produce a Markovian short rate process. Of course, if we choose the initial forward rate curve r^* as $r^* = \hat{r}$, then we come back where we started. However, if we choose another initial forward rate curve r^* (for example the observed forward rate curve), then it is no longer clear if the short rate will be Markovian. Theorem 5.7.4 says that we will obtain, for all initial points r^* in a neighborhood of $\hat{r}$, a Markovian short rate model, if and only if the original model is affine.

References

[1] BJÖRK T., "Arbitrage Theory in Continuous Time", Oxford University Press, 1998.

[2] BJÖRK T. – CHRISTENSEN B. J., *Interest rate dynamics and consistent forward rate curves*, Mathematical Finance, **9** no. 4 (1999), 323-348.

[3] BJÖRK T. – GOMBANI A., *Minimal realization of interest rate models*, Finance and Stochastics, **3** no. 4 (1999), 413-432.

[4] BJÖRK T. – SVENSSON L., *On the existence of finite dimensional realizations for nonlinear forward rate models*, to appear in Mathematical Finance, 1999.

[5] BROCKETT, R.W., "Finite Dimensional Linear Systems", Wiley, 1970.

[6] DA PRATO G. – ZABCZYK J., "Stochastic Equations in Infinite Dimensions", Cambridge University Press, 1992.

[7] DELBAEN F. – SCHACHERMAYER W., *A general version of the fundamental theorem on asset pricing*, Mathematishe Annalen, **300** (1994), 463-520.

[8] FILIPOVIĈ D., *A Note on the Nelson-Siegel family*, Mathematical Finance, **9** no. 4 (1998a), 349-359.

[9] FILIPOVIĈ D., *Exponential-polynomial families and the term structure of interest rates*, to appear in Bernoulli (1998b).

[10] JACOD J. – SHIRYAEV A.N., "Limit Theorems for Stochastic Processes", Springer-Verlag, Berlin, 1987.

[11] JEFFREY A., *Single factor Heath-Jarrow-Morton term structure models based on Markovian spot interest rates*, JFQA, **30** no. 4 (1995), 619-642.

[12] NELSON, C. – SIEGEL, A., *Parsimonious modelling of yield curves*, Journal of Business, **60** (1987), 473-489.

PUBBLICAZIONI DELLA CLASSE DI SCIENZE DELLA SCUOLA NORMALE SUPERIORE

QUADERNI

1. DE GIORGI E., COLOMBINI F., PICCININI L.C.: *Frontiere orientate di misura minima e questioni collegate.*
2. MIRANDA C.: *Su alcuni problemi di geometria differenziale in grande per gli ovaloidi.*
3. PRODI G., AMBROSETTI A.: *Analisi non lineare.*
4. MIRANDA C.: *Problemi in analisi funzionale* (ristampa).
5. TODOROV I.T., MINTCHEV M., PETKOVA V.B.: *Conformal Invariance in Quantum Field Theory.*
6. ANDREOTTI A., NACINOVICH M.: *Analytic Convexity and the Principle of Phragmén-Lindelöf.*
7. CAMPANATO S.: *Sistemi ellittici in forma divergenza. Regolarità all'interno.*
8. *TOPICS IN FUNCTIONAL ANALYSIS: Contributors:* F. STROCCHI, E. ZARANTONELLO, E. DE GIORGI, G. DAL MASO, L. MODICA.
9. LETTA G.: *Martingales et intégration stochastique.*
10. *OLD AND NEW PROBLEMS IN FUNDAMENTAL PHYSICS: Meeting in honour of* G.C. WICK.
11. *INTERACTION OF RADIATION WITH MATTER: A Volume in honour of* ADRIANO GOZZINI.
12. MÉTIVIER M.: *Stochastic Partial Differential Equations in Infinite Dimensional Spaces.*
13. *SYMMETRY IN NATURE: A Volume in honour of* LUIGI A. RADICATI DI BROZOLO.
14. *NONLINEAR ANALYSIS: A Tribute in honour of* GIOVANNI PRODI.
15. LAURENT-THIÉBAUT C., LEITERER J.: *Andreotti-Grauert Theory on Real Hypersurfaces.*
16. ZABCZYK J.: *Chance and Decision. Stochastic Control in Discrete Time.*
17. EKELAND I.: *Exterior Differential Calculus and Applications to Economic Theory.*

CATTEDRA GALILEIANA

1. LIONS P.L.: *On Euler Equations and Statistical Physics.*
2. BJÖRK T.: *A Geometric View of the Term Structure of Interest Rates.*

LEZIONI LAGRANGE

1. VOISIN C.: *Variations of Hodge Structure of Calabi-Yau Threefolds.*

LEZIONI FERMIANE

1. THOM R.: *Modèles mathématiques de la morphogénèse.*
2. AGMON S.: *Spectral Properties of Schrödinger Operators and Scattering Theory.*
3. ATIYAH M.F.: *Geometry of Yang-Mills Fields.*
4. KAC M.: *Integration in Function Spaces and Some of Its Applications.*
5. MOSER J.: *Integrable Hamiltonian Systems and Spectral Theory.*
6. KATO T.: *Abstract Differential Equations and Nonlinear Mixed Problems.*
7. FLEMING W.H.: *Controlled Markov Processes and Viscosity Solution of Nonlinear Evolution Equations.*
8. ARNOLD V.I.: *The Theory of Singularities and Its Applications.*
9. OSTRIKER J.P.: *Development of Larger-Scale Structure in the Universe.*
10. NOVIKOV S.P.: *Solitons and Geometry.*
11. CAFFARELLI L.A.: *The Obstacle Problem.*

ALTRE PUBBLICAZIONI

Proceedings of the Symposium on FRONTIER PROBLEMS IN HIGH ENERGY PHYSICS
Pisa, June 1976

Proceedings of International Conferences on SEVERAL COMPLEX VARIABLES, Cortona,
June 1976 and July 1977

*Raccolta degli scritti dedicati a JEAN LERAY apparsi sugli Annali della Scuola Normale
Superiore di Pisa*

*Raccolta degli scritti dedicati a HANS LEWY apparsi sugli Annali della Scuola Normale
Superiore di Pisa*

*Indice degli articoli apparsi nelle Serie I, II e III degli Annali della Scuola Normale
Superiore di Pisa* (dal 1871 al 1973)

*Indice degli articoli apparsi nella Serie IV degli Annali della Scuola Normale Superiore
di Pisa* (dal 1974 al 1990)

ANDREOTTI A.: *SELECTA vol. I, Geometria algebrica.*

ANDREOTTI A.: *SELECTA vol. II, Analisi complessa, Tomo I e II.*

ANDREOTTI A.: *SELECTA vol. III, Complessi di operatori differenziali.*

Fotocomposizione "CompoMat" Loc. Braccone, 02040 Configni (RI), Italy
Finito di stampare per conto della "CompoMat" dalla Nuova Grafica 86 nel settembre 2001